T0305859

Bioinspired Materials and Metamaterials

Development of bioinspired materials and metamaterials has changed the philosophy of materials engineering and opened new technological possibilities, as they demonstrate properties that are not found in naturally occurring materials. This book examines advances in these emerging materials classes and investigates how their tailor-engineered properties, such as specific surface energy or refraction index, enable the design of devices and ultimately the ability to solve complex societal problems that are, in principle, impossible with traditional materials.

The aim of this book is to survey the scientific foundations of the design and properties of bioinspired materials and metamaterials and the way they enter engineering applications.

- Introduces the physico-chemical foundations, theoretical groundings, and main equations of biomimetic and metamaterials science
- Describes how to develop and design these advanced materials and their applications
- Features end-of-chapter problems to help readers apply the principles
- Surveys achievements, including metamaterials cloaking and the negative mass effect
- Emphasizes ecological aspects of materials science

The text is intended for materials engineering students who have completed courses in general physics, chemistry, and calculus, as well as researchers in materials science and engineering.

Emerging Materials and Technologies

Series Editor: Boris I. Kharissov

The *Emerging Materials and Technologies* series is devoted to highlighting publications centered on emerging advanced materials and novel technologies. Attention is paid to those newly discovered or applied materials with potential to solve pressing societal problems and improve quality of life, corresponding to environmental protection, medicine, communications, energy, transportation, advanced manufacturing, and related areas.

The series takes into account that, under present strong demands for energy, material, and cost savings, as well as heavy contamination problems and worldwide pandemic conditions, the area of emerging materials and related scalable technologies is a highly interdisciplinary field, with the need for researchers, professionals, and academics across the spectrum of engineering and technological disciplines. The main objective of this book series is to attract more attention to these materials and technologies and invite conversation among the international R&D community.

Advanced Synthesis and Medical Applications of Calcium Phosphates
Edited by S.S. Nanda, Jitendra Pal Singh, Sanjeev Gautam, and Dong Kee Yi

Non-Metallic Technical Textiles: Materials and Technologies
Mukesh Kumar Sinha and Ritu Pandey

Smart Micro- and Nanomaterials for Drug Delivery
Edited by Ajit Behera, Arpan Kumar Nayak, Ranjan K. Mohapatra, and Ali Ahmed Rabaan

Smart Micro- and Nanomaterials for Pharmaceutical Applications
Edited by Ajit Behera, Arpan Kumar Nayak, Ranjan K. Mohapatra, and Ali Ahmed Rabaan

Friction Stir-Spot Welding: Metallurgical, Mechanical and Tribological Properties
Edited by Jeyaprakash Natarajan and K. Anton Savio Lewise

Phase Change Materials for Thermal Energy Management and Storage: Fundamentals and Applications
Edited by Hafiz Muhammad Ali

Nanofluids: Fundamentals, Applications, and Challenges
Shriram S. Sonawane and Parag P. Thakur

MXenes: From Research to Emerging Applications
Edited by Subhendu Chakroborty

Biodegradable Polymers, Blends and Biocomposites: Trends and Applications
Edited by A. Arun, Kunyu Zhang, Sudhakar Muniyasamy and Rathinam Raja

Bioinspired Materials and Metamaterials: A New Look at the Materials Science
Edward Bormashenko

For more information about this series, please visit: www.routledge.com/Emerging-Materials-and-Technologies/book-series/CRCEMT

Bioinspired Materials and Metamaterials

A New Look at the Materials Science

Edward Bormashenko

CRC Press
Taylor & Francis Group
Boca Raton London New York

CRC Press is an imprint of the
Taylor & Francis Group, an **informa** business

First edition published 2025
by CRC Press
2385 NW Executive Center Drive, Suite 320, Boca Raton FL 33431

and by CRC Press
4 Park Square, Milton Park, Abingdon, Oxon, OX14 4RN

CRC Press is an imprint of Taylor & Francis Group, LLC

ISBN: 978-1-032-01403-6 (hbk)
ISBN: 978-1-032-01404-3 (pbk)
ISBN: 978-1-003-17847-7 (ebk)

DOI: 10.1201/9781003178477

Typeset in Times
by KnowledgeWorks Global Ltd.

To my beloved wife

Contents

About the Author

Edward Bormashenko is Professor of Materials Science and the Head of the Laboratory of Interface Science at Ariel University in Israel. He was born in 1962 in Kharkiv, Ukraine, and has lived in Israel since 1997. He studied at the V. N. Karazin Kharkiv National University. His research is in polymer science and surface science. He received his PhD (supervised by Professor M. L. Friedman) from Moscow Institute of Plastics in 1990. His scientific interests include metamaterials, superhydrophobicity, superoleophobicity, creating of surfaces with prescribed properties, plasma- and UV-treatment of surfaces, plasma treatment of seeds, liquid marbles and their self-propulsion and the Moses effect (magnetically inspired deformation of liquid surfaces). Professor Bormashenko is also active in quantitative linguistics, topological problems of physics (exemplifications of the "hairy ball" theorem), advanced dimensional analysis (extensions of the Buckingham theorem), variational analysis of "free ends" physical problems, enabling application of the "transversality conditions". He is one of the most productive and cited scientists at Ariel University and is the author of three monographs.

Preface

HISTORICAL BACKGROUND

Etymology of the word "material" stems from Late Latin māteriālis as well as from Latin māteria ("wood, material, substance"). Perhaps, the most surprising linguistic finding reveals that this etymology arises from māter ("mother"). Indeed, in a certain sense, the materials science is a mother of engineering and technology. Nothing may be produced without materials. Actually, the lion's share of the overall technological progress is due to the progress in the materials science and engineering. Certain ancient periods of history are named after the material that was predominantly utilized at that time. The Stone age, Bronze age, Iron age and the Polymer age reflect stages of development of materials science and engineering. It is hard to predict how the modern era will be named, but it is possible that it will be labeled as the Metamaterials age. Our book is devoted to the evergreen materials science, more accurately speaking, to its newborn branches: bioinspired materials and metamaterials.

WHY ONE MORE BOOK ON THE MATERIALS SCIENCE? WHAT IS DIFFERENT ABOUT THIS BOOK?

Dozens of excellent books, devoted to the materials science, were published. Some of them may be definitely recommended to BSc and MSc students studying the materials science. More advanced courses are appropriate for engineers and experts in the field of the materials science. So, why one more book devoted to the materials science is suggested? The present book is not intended to replace traditional courses surveying the materials science. It is focused on the breakthrough attained in the materials science, during last decades, namely rapid, explosive development of biomimetic materials and metamaterials. A glance at the content pages will show sections dealing with:

The "lotus effect" and superhydrophobic materials
Rose petal effect
Salvinia effect
Shark-skin effect
Gecko effect and novel adhesives
Negative refractive index and related metamaterials, double-negative materials
Acoustic metamaterials
Negative bulk modulus and negative Poisson's ratio materials

All of aforementioned chapters deal with the materials of 21st century, which are expected to replace the traditional metallic, polymer and ceramic-based solutions. What is common for bioinspired and metamaterials? They are materials with *prescribed, tailor-designed properties*, such as specific surface energy or refraction index. Development of bioinspired materials and metamaterials changed the philosophy of materials engineering and opened new technological possibilities inaccessible to the traditional materials science engineering. This switch in the engineering thinking became possible due to the fact that recently developed artificial materials demonstrate properties that are not found in naturally occurring materials. The impact of metamaterials and biomimetic materials is well-expected to be enormous. If one can tailor and manipulate the properties of materials, significant decrease in the size and weight of components, devices and systems along with enhancements in their appearance appears to be realizable. Moreover, materials with tailor-engineered properties enable design of the devices, which are, in principle, impossible with traditional materials.

A number of books already appeared, which may be recommended to the students and engineers specialized in the materials science. However, the proposed book is the first textbook in the field, supplying to a reader a broad acquaintance with a profoundly scientifically grounded progress in the novel materials science. The aim of this book is to present new materials, to survey deeply scientific foundations of their design and properties and the way they enter engineering applications.

HOW TO READ THIS BOOK?

The text is intended for engineering students who have completed courses in general physics, chemistry and calculus. To facilitate understanding, a brief description of key ideas from earlier mandatory courses is provided to help students refresh their memories. Based on these ideas, the book introduces first the physico-chemical foundations of bioinspired materials and metamaterials. The main notions and equations of the biomimetic and metamaterials science, such as the refraction index, the surface tension, the contact angle, the Young equation, the Cassie-Baxter and Wenzel wetting models and Maxwell equations, are treated in much detail. Based on this foundation, the theoretical groundings of the bioinspired materials and metamaterials are developed. Numerous applications of biomimetic and metamaterials engineering are discussed, making the presentation practice-oriented and useful for both students and materials engineers. Numerous end-of-chapter problems are prepared to give the student practice in applying the principles, presented in each chapter.

I am thankful to my parents who instilled in me at an early age an interest to scientific research. I am in particular indebted to my scientific mentor, Professor of Kharkov University, Yakov Evseevitch Gegusin for his inspiring investigations in the field of wetting phenomena. I am grateful to my PostDoc and PhD students Irina Legchenkova and Artem Gilevich for their inestimable help in preparing this book. I am grateful to Dr. Mark Frenkel for his longstanding fruitful cooperation in the study of bioinspired materials and metamaterials. I am especially indebted to my wife Yelena Bormashenko for her inestimable help in preparing this book.

Symbol Index

ε	the dielectric *permittivity*
ε_0	absolute dielectric vacuum permittivity
η	viscosity
θ	angle
θ_D	dynamic contact angle
θ_Y	Young contact angle
μ	magnetic permeability
μ_0	magnetic permeability of vacuum
ν	Poisson ratio (Section 10.3)
$\tilde{\upsilon}_{kin}$	kinematic viscosity
Π	disjoining pressure
ρ	density
τ	propagation time (Section 8.5)
σ	electrical conductivity
ω_p	plasma frequency
Ψ	spreading parameter

1 Bioinspired Materials
Introduction to Main Physical Notions—Surface Tension and Viscosity

In this chapter, we introduce biomimetic materials and their classification. We will also introduce and discuss in detail main physical notions, such as surface tension and viscosity, which are crucial for understanding of properties of biomimetic materials; afterwards, we will address foundations of the modern interface surface. We address the temperature dependencies of surface tension and viscosity of liquids and their physical origin. We also consider surface tension of solids.

1.1 WHAT ARE BIOINSPIRED MATERIALS? CLASSIFICATION OF BIOMIMETIC MATERIALS

The notion "biomimetics" stems from the Greek word "biomimesis". It means mimicking biology or living nature, or living organisms, and is also called biomimicry.[1] Biomimetic materials are also called bioinspired ones. Biological materials are usually hierarchically built when comprising well-developed patterns appearing on various scales: starting from the molecular via the nano- and micro- to the macroscale.[1] The field of biomimetics or bioinspired hierarchically structured surfaces started to progress explosively in the early 2000s when technological achievements in micro- and nano-technologies enabled industrial manufacturing of micro- and nano-reliefs. When we speak about biomimetic materials, we mean mainly biomimetic surfaces; thus, today the field of biomimetic materials appears mainly as a field of the interface science.[2–4]

Through the long history of natural evolution, many creatures have adapted themselves to the surrounding environments and evolved remarkable ability to interact with water. Biological aspects of biomimetics are surveyed in Reference 5. We focus on the materials science-related aspects of development of biomimetic materials. The physical properties of a solid surface depend strongly on its chemical compositions on the one hand and on its topography on the other hand.[1,3,4,6,7] Most biological materials take the strategy of hierarchical structures to achieve their multiple surface functions, e.g., superhydrophobicity, water collection and transport. Interrelations between the biological functions, physical properties, chemical compositions and geometric structures of biological materials were studied intensively recently.[1,4,6]

Various classification schemes describing the realm of bioinspired materials were suggested. Biomimetic surfaces may be classified according to their functionality

DOI: 10.1201/9781003178477-1

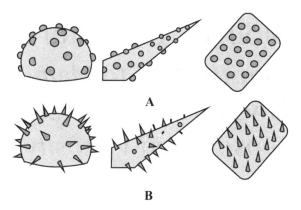

FIGURE 1.1 (**A**) The upper schemes represent 0D topographies built of 0D "spots". (**B**) The lower scheme shows typical 1D topographies comprising 1D needles.

and thus should be divided into self-cleaning, icephobic, anti-corrosive and drag reduction ones. According to this scheme, the lotus-leaf-like biomimetic surfaces are mainly self-cleaning ones; surfaces mimicking Gecko feet are exploited as adhesives, whereas *Salvinia-* and shark-skin-inspired surfaces are used for the drag reduction.[1,8] An alternative classification scheme based on the dimensionality of elements constituting the surface was suggested in Reference 6. This classification considers the morphological dimensionality of structure units at different length scales.[6] The structure units on each level are classified into 0D, 1D and 2D according to their prominent geometric features. According to this classification scheme, the lotus-leaf-mimicking surfaces are related to 0D interfaces, due to the point-like nature of elements constituting its topography (see Figure 1.1A).[6]

The hairy legs of water striders, providing their remarkable ability to walk along water, are related to 1D materials (see Figure 1.1B).[6] Water strider leg is covered by thousands of microscopically hydrophobic needle-like setae, which may be considered as 1D elements.[6] Butterfly wings demonstrating remarkable anisotropic wetting are built, in turn, from 2D topographical elements.[6] The usefulness of the classification scheme as always depends on the prescribed goal of the classification: whether scientific or engineering.

1.2 SURFACE TENSION AND ITS DEFINITION

Now let us acquaint the notion of the surface tension, which is crucial for understanding of physics and engineering of biomimetic surfaces. Surface tension is one of the most fundamental properties of liquid and solid phases. Actually condensed phases (liquid and solid) differ from the gaseous phase in that they are bounded by their own surface, which may be seen as an inevitable defect inherent for solids and liquids. Surface tension governs a diversity of natural and technological effects, including floating of a steel needle, capillary rise, walking of water striders on the water surface, washing and painting. It governs many phenomena in climate formation, plant biology and medicine. Surface tension is exactly what it says: the tension

FIGURE 1.2 Manifestation of surface tension: steel needle floating on a water surface.

in a surface and the reality of its existence are demonstrated in Figure 1.2, presenting a metallic needle supported by a water surface. We will discuss surface-tension-supported floating in detail below; now let us define exactly the notion of the surface tension and its quantitative interpretation.

Imagine a rectangular metallic frame closed by a mobile piece of wire as depicted in Figure 1.3. If one deposits a soap film within the rectangle, the film will want to diminish its surface area. Thus, it acts perpendicularly and uniformly on the mobile wire as shown in Figure 1.3. The surface tension $\vec{\gamma}$ could be defined as a force per unit length of the wire; thus, the dimension of the surface tension is $[\gamma] = \frac{N}{m}$.

The surface tension defined in this way is a tensor which acts perpendicularly to a line in the surface. Surface tension is often identified with the specific surface free energy. Indeed, when the mobile rod l in Figure 1.3 moves by a distance dx, the work $2\gamma l dx$ is done (the factor of "2" reflects the presence of upper and lower interfaces). Thus, the surface tension γ could be identified with the energy supplied to increase

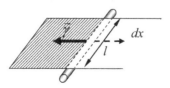

FIGURE 1.3 The definition of surface tension: *force* normal to the line (rod).

TABLE 1.1

Surface Tension, Enthalpy of Vaporization and Dipole Moment of Some of Organic Molecules

Liquid	Surface Tension (Ambient Conditions), (mJ/m²)	Molar Enthalpy of Vaporization, (ΔH_{mol}, kJ/mol)	Dipole Moment, $\tilde{p}$ (D*)
Glycerol, $C_3H_8O_3$	64.7	91.7	2.56
Formamide, CH_3ON	55.5	60.0	3.7
CCl_4	25.7	32.54	0
Chloroform, $CHCl_3$	26.2	31.4	1.04
Dichloromethane, CH_2Cl_2	31	28.6	1.60
Toluene, C_7H_8	28.5	38.06	0.36
Ethyl alcohol, C_2H_6O	22	38.56	1.7
Acetone, C_3H_6O	24	31.3	2.9

* The unit of a dipole moment is Debye: $1D = 3.3 \cdot 10^{-30} C \cdot m$.

the surface area by one unit (consequently, the dimension of the surface tension is alternatively expressed as $[\gamma] = \frac{J}{m^2}$).

This identification may give rise to misinterpretations: the surface tension defined as force per unit length of a line in the surface is a *tensor*, whereas specific surface free energy is a *scalar* thermodynamic property of an area of the surface without directional attributes.[3,9] However, for liquids at a constant temperature and pressure and in equilibrium, the surface tension is numerically equal and physically equivalent to the specific surface Helmholtz free energy.[3,9] Let us start from this simplest situation, i.e., the surface tension of liquids in equilibrium. Surface tension of water at the temperature $t = 25 \,°C$ equals $\gamma = 71.97 \pm 0.05 \frac{mJ}{m^2}$; surface tensions of a number of organic liquids are supplied in Table 1.1. It is easily recognized that the surface tension of water is unusually high and it is higher than that of organic liquids. It should be emphasized that the surface tension is an essentially macroscopic property of the medium. Now, we will try to understand how the surface tension emerges from the peculiarities of interactions between the molecules constituting the medium. *So we acquaint extremely important concept of the materials science: the problems may be seen macroscopically (at the scale of the entire physical body or medium) or microscopically (at the level of molecules, constituting the medium).*

1.3 PHYSICAL ORIGIN OF THE SURFACE TENSION OF MONO-COMPONENT LIQUIDS

Liquid is a condensed phase in which molecules interact. The origin of surface tension is related to the unusual energetic state of the surface molecule, which misses half its interactions (see Figure 1.4). The energy states of molecules in the bulk and at the surface of liquid are not the same due to the difference in the nearest surrounding of a given molecule. Each molecule in the bulk is surrounded by others on every

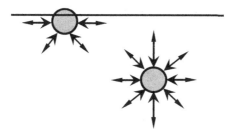

FIGURE 1.4 A molecule at the surface misses about half its interactions.

side, whereas for the molecule located at the liquid/vapor interface, there are very few molecules outside of the liquid, as shown in Figure 1.4.

Here, a widespread misinterpretation should be avoided; the resulting force acting on the molecule in the bulk and at the interface equals zero (both "bulk" and "interface" molecules are in mechanical equilibrium). For example, we can read: "the unbalanced force on a molecule is directed inward".[2] If this is the case, the molecule according to Newton's Second Law has to move toward the bulk, and all the liquid has to flow instantaneously in obvious conflict with the energy conservation. This common misinterpretation was revealed and carefully analyzed in Reference 10. Let us clarify the situation. Figure 1.5, depicting an "instantaneous photo" of the potential relief, describing the interaction of a molecule of liquid with its surrounding, clarifies the situation. If all molecules are supposed to be fixed, the potential energy of a molecule will change, as shown schematically in Figure 1.5. Obviously, the total force acting on a molecule in equilibrium is zero.

However, an increase in the liquid/vapor surface causes a rise in the quantity of "interface" molecules and a consequent growth in the surface energy. Liquids tend to diminish the number of interface molecules to decrease surface energy. Thus, the surface free energy of the material is the work that should be supplied to bring the molecules from the interior bulk phase to its surface to create a new surface having

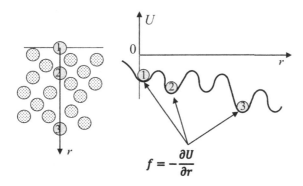

FIGURE 1.5 A potential relief describing the interaction of a molecule of liquid with its surrounding.

a unit area. Let the potential describing the pair intermolecular interaction in the liquid be $U(r)$. The surface tension γ could be estimated as follows:

$$\gamma = f_m \frac{1}{d_m} \cong \frac{N}{2} \frac{|U(d_m)|}{d_m} \frac{1}{d_m} = \frac{N}{2} \frac{|U(d_m)|}{d_m^2}, \tag{1.1}$$

where f_m is the force necessary to bring a molecule to the surface, which could be roughly estimated as $f_m \cong \frac{N}{2} \frac{|U(d_m)|}{d_m}$, where d_m is the diameter of the molecule, N is the number of nearest-neighbor molecules (the multiplier $1/2$ is due to the absence of molecules "outside", i.e., in the vapor phase) and $1/d_m$ is the number of molecules per unit length of the liquid surface. The diameter of the molecule may be estimated as $d_m \cong \lambda a_0$, where $a_0 = \frac{h^2}{me^2}$ is the Bohr radius (m and e are the mass and charge of the electron, respectively, and h is Planck's constant) and $6 < \lambda < 12$.[9,10] It is seen from Eq. (1.1) that the surface tension in liquids is defined by the pair intermolecular interaction $U(r)$, to be discussed in detail in Section 1.5; 1, the diameter of the molecule d_m and the number N. Now let us cast a glance on Table 1.1, supplying surface tensions of a number of liquids. The similar values of surface tensions, summarized in Table 1.1, of liquids which are very different in their physical and chemical nature catch the eye. Indeed, the values of surface tension of most of organic liquids are located in the narrow range of 20–65 mJ/m².

The proximity of surface tension of most of organic liquids deserves the detailed discussion to be supplied below (see Section 6.2.3; actually, the proximity of the surface tension of a majority of organic liquids is due to the dominating nature of the London (dispersion) forces in the van der Waals interactions); however, let us acquaint the second important property of liquids, namely its viscosity.

1.4 VISCOSITY OF LIQUIDS

Viscosity is a measure of a fluid's resistance to flow. It describes the internal friction of a moving fluid. A fluid with large viscosity resists motion because its molecular makeup gives it a lot of internal friction. A fluid with low viscosity flows easily because its molecular makeup results in very little friction when it is in motion. Consider viscous flow in a liquid confined between two parallel plates, as shown in Figure 1.6.

Formally, viscosity (represented by the symbol η "eta") is the ratio of the shearing stress $\tau = \frac{F}{A}$ to the velocity gradient $\frac{\partial u}{\partial y}$ in a fluid, where F is the force, A area and u velocity (see Figure 1.6):

$$\eta = \frac{\tau}{\partial u/\partial y} = \frac{F/A}{\partial u/\partial y} \tag{1.2}$$

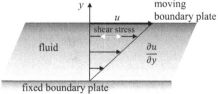

FIGURE 1.6   Scheme illustrating viscous flow of liquids.

It is convenient to re-shape Eq. (1.2) in the form of the Newton equation, which states that the shear stress τ is directly proportional to the viscosity η and gradient of velocity $\frac{\partial u}{\partial y}$, namely:

$$\tau = \eta \frac{\partial u}{\partial y} \tag{1.3}$$

Transport theory provides an alternative interpretation of viscosity in terms of momentum transport: viscosity is the material property which characterizes momentum transport within a fluid or gas. The SI unit of viscosity is the Pascal second $[Pa \times s]$, which has no special name, $[\eta] = Pa \times s = \frac{kg}{m \times s}$. The ubiquitous CGS unit of viscosity is poise, denoted P; conversion from system of units to another is according to $10P = 1Pa \times s$. Viscosity (as well as surface tension) is essentially a macroscopic property of the medium.

Viscosity of common liquids is strongly (exponentially) temperature-dependent and described by the Andrade equation (see References 11 and 12):

$$\eta(T) = \tilde{\eta} exp \frac{E_a}{k_B T} = \tilde{\eta} exp \frac{E_{amol}}{RT}, \tag{1.4}$$

where $\tilde{\eta}$ is a constant, k_B is the Boltzmann constant, R is the gas constant, E_a and E_{amol} are the specific and molar energies of activation of viscous flow addressed by Eyring et al. in Reference 11; for water, $E_{amol} \cong 15.75 \frac{kJ}{mol}$ (energies of activation of viscous flow for a broad variety of liquids are summarized in Reference 13). Eq. (1.4) is obviously an Arrhenius-like one. Eyring considered viscosity as a rate process, namely shear viscosity in a liquid involves an exchange of position between a hole and its adjacent molecule and, thus, justified Eq. (1.4).[11] Viscosity of glass-forming liquids is described by the Vogel-Fulcher-Tammann equation:

$$\eta(T) = \eta_0 exp \frac{B}{T - T_{VFT}}, \tag{1.5}$$

or, alternatively, Eq. (1.5) may be represented in the logarithmic form, as follows:

$$log_{10}\eta(T) = log_{10}\eta_0 + \frac{A}{T - T_{VFT}} \tag{1.6}$$

where η_0, B and $A = Blog_{10}e$ are empirical liquid-dependent parameters and T_{VFT} is also an empirical fitting parameter, which lies below the glass transition temperature.[14,15] These three parameters are normally used as adjustable parameters to fit the Vogel-Fulcher-Tammann equation to experimental data of specific systems. The values of the fitting parameters of the Vogel-Fulcher-Tammann equation for some simple liquids are supplied in Table 1.2 (the precise values of these parameters for a diversity of glass-forming liquids are supplied in Reference 15).

The following fact should be emphasized: the viscosity of certain liquids can increase by up to 13 orders of magnitude within a relatively narrow temperature interval, and this is due to the very strong exponential temperature dependence of viscosity, described by Eqs. (1.4)–(1.6). Moreover, viscosities of various liquids at

TABLE 1.2

Fitting Parameters of the Vogel-Fulcher-Tammann Equation for Some of Simple Liquids

Liquid	A (K)	T_{VFT} (K)
SiO_2	21.254	139
B_2O_3	4.695	252

the same temperature also differ very strongly. For example, the viscosity of ethyl alcohol at ambient conditions equals $1.2 \cdot 10^{-3}$ kg/m·s, whereas the viscosity of glycerol is 1.5 kg/m·s and this is in a striking contrast to the surface tensions of liquids: we already mentioned in the previous section that the surface tensions of alcohol and glycerol are of the same order of magnitude. Even more striking example is honey, the viscosity of which may be very high; however, its surface tension is 50–60 mJ/m² and it is slightly lower than that of water. The reasonable question is: why is the range of values of surface tension so narrow? We address this question in the next section.

1.5 VISCOSITY OF LIQUIDS *VS.* SURFACE TENSION

Now, we adopt the *microscopic approach,* and we'll try to understand how viscosity arises from the intermolecular interaction. Surface tension of liquids to a much extent is defined by the intermolecular potential $U(r)$ as demonstrated by Eq. (1.1). In general, there are three main kinds of intermolecular interactions:

1. The attractive interaction between identical dipolar molecules (embedded into the medium with the dielectric constant), given by the Keesom formula:

$$U_K(r) = -\frac{\tilde{p}^4}{3(4\pi\varepsilon\varepsilon_0)^2 k_B T} \frac{1}{r^6},\qquad(1.7)$$

where $\tilde{p}$ is the dipole moment of the molecule, k_B is the Boltzmann constant, T is the temperature, ε_0 is the vacuum permeability and r is the distance between molecules.[3,16]

2. The Debye attractive interaction between dipolar molecules and induced dipolar molecules is

$$U_D(r) = -\frac{2\tilde{p}^2\alpha}{(4\pi\varepsilon\varepsilon_0)^2} \frac{1}{r^6},\qquad(1.8)$$

where α is the polarizability of the molecule (see also Section 6.2.2).[3,16]

3. The London dispersion interactions which are of a pure quantum mechanical nature. The London dispersion force is an attractive force that results when the electrons in two adjacent atoms occupy positions that make the atoms form temporary dipoles; its potential is given by

$$U_L(r) = -\frac{3\alpha^2 I}{4(4\pi\varepsilon_0)^2} \frac{1}{r^6}, \qquad (1.9)$$

where I is the ionization potential of the molecule.[3,16] All attractive intermolecular interactions given by Formulas (1.7)–(1.9) decrease as $\frac{1}{r^6}$. The importance of the power law index -6 is discussed **in Appendix 1.A**. The physical origin of these interactions will be discussed in detail in Sections 6.2.2 and 6.2.3 in their relation to the Gecko effect, and Gecko-effect-inspired dry adhesives.

The Keesom, Debye and London interactions are collectively termed van der Waals interactions. It should be stressed that the London dispersion forces given by Formula (1.9) governs intermolecular van der Waals interactions in most of organic liquids. They are several orders of magnitude larger than the dipole-dipole Keesom and Debye forces described by Expressions (1.7) and (1.8).[16,17] Taking this into account, we obtain with Formulas (1.1) and (1.9) a very simple (and crude) estimation of the surface tension of liquids (for details, see Reference 17):

$$\gamma \cong \frac{3N}{2^{10}} \frac{I}{d_m^2}. \qquad (1.10)$$

Formula (1.10) answers the question: why do surface tensions of most organic liquids demonstrate close values? Indeed, it is seen from (1.10) that the surface tension of a broad variety of organic liquids depends on the potential of the ionization and the diameter of the molecule only. These parameters vary slightly for all organic liquids. Formula (1.10) predicts for simple liquids a surface tension roughly close to the values displayed in Table 1.1.[17] Moreover, Formula (1.10) predicts $\gamma \sim \frac{const.}{d_m^2}$; this dependence actually takes place for n-alkanes.[18] Moreover, molar enthalpies of vaporization (supplied in Table 1.1) and tensile strengths of most liquids (which are also governed by intermolecular forces) are of the same order of magnitude. On the contrary, *viscosity is an activation phenomenon*, demanding surmounting of a potential barrier, when a molecule of liquid jumps from one equilibrium position to another; thus, its temperature dependence quite naturally is supplied by the Arrhenius-like Eqs. (1.4)–(1.6).

The London dispersion force will dictate the surface tension of a liquid when hydrogen or metallic (mercury) bonds acting between molecules could be neglected. When hydrogen or metallic bonds are not negligible it was supposed that the surface tension of liquids could be presented in an additive way:

$$\gamma = \gamma^d + \gamma^h; \gamma = \gamma^d + \gamma^{met}, \qquad (1.11)$$

where the first term γ^d represents the dispersion London force contribution and the second term represents the hydrogen (labeled in Eq. (1.11) with γ^h) or metallic (denoted γ^{met}) bonding.[19] However, the concept of additivity of surface tension components was criticized by several groups, and it was shown that there exist liquids for which Eq. (1.11) becomes problematic.[20]

1.6 TEMPERATURE DEPENDENCE OF THE SURFACE TENSION

When the temperature is increased, the kinetic agitation of the molecules increases. Thus, the molecular interactions become more and more weak compared to the kinetic energy of the molecular motion. Hence, it is quite expectable that the surface tension will decrease with the temperature. The temperature dependence of the surface tension is well described by the Eötvös equation (Eötvös rule):

$$(V_{ML})^{2/3} \gamma = \hat{k}(T_c - T), \tag{1.12}$$

where V_{ML} is the molar volume of the liquid: $V_{ML} = M_W/\rho_L$, M_W and ρ_L are the molar mass and the liquid density, respectively, T_c is the critical temperature of a liquid and $\hat{k}$ is a constant valid for all liquids. The Eötvös constant has a value of $\hat{k} = 2.1 \times 10^{-7}$ J/mol$^{2/3}$K. An abundance of modifications of the Eötvös formula (1.12) has been proposed; however, for practical purposes, the linear dependence of the surface tension could be supposed.[3,9]

1.7 IMPACT OF ENTROPY ON THE SURFACE TENSION: THE EFFECT OF SURFACE FREEZING

Until now we discussed the surface tension as a pure energetic effect. Actually, the situation is more complicated and an impact of entropy on the surface tension should be considered. The surface tension γ as a direct measure of the surface excess free energy is given by

$$\gamma = W_S - W_B - T(S_S - S_B), \tag{1.13}$$

where W_S and W_B are the energies and S_S and S_B the entropies for the surface and bulk, respectively. The temperature slope of surface tension yields information on the surface excess entropy: $\frac{d\gamma}{dT} = -(S_S - S_B)$, which is directly related to the ordering and disordering of molecules on the surface. For ordinary liquid surfaces, the molecules on the surfaces are less constrained than those in bulk; thus, S_S is slightly larger than S_B, yielding $\frac{d\gamma}{dT} < 0$, as it is predicted by the Eötvös equation (Eq. (1.12)). A negative temperature slope has indeed been observed for all the simple liquids. However, for normal alkanes and some other medium-sized linear hydrocarbons, a positive temperature slope has been observed in a small temperature region above their melting point, indicating a higher ordering of molecules at the surface, comparatively to the bulk. This rare effect was called "the surface freezing".[21,22]

It is recognized from Eq. (1.13) that the surface tension could not be identified with the energy of the unit area of a liquid. The rigorous thermodynamic definition of the surface tension involves Helmholtz or Gibbs free energies, including entropy contribution (see **Appendix 1.B**).

It should be emphasized that the linear dependence of surface tension summarized by the Eötvös equation (1.12) is much weaker than the exponential temperature dependence of viscosity, described by Eqs. (1.4) and (1.5) (we already discussed this difference in Sections 1.4 and 1.5). The question is what is the physical reasoning for this difference? The explanation of this observation is based on the fact that viscosity is a kind of activation phenomenon, necessarily, including surmounting of the potential barrier, as suggested by Eyring et al. in Reference 11, whereas the temperature dependence of the surface tension emerges as mainly an entropic effect (see Eq. (1.13)).

1.8 SURFACTANTS

Surface tension of liquids could be modified not only physically but also chemically by introducing surfactants. A surfactant is a molecule which has two parts with different affinities. One of these parts has an affinity to non-polar media and the second part has an affinity to polar media such as water. The energetically most favorable orientation for these molecules may be attained at surfaces or interfaces so that each part of the molecule can reside in an environment for which it has the greatest affinity.

In most cases, the hydrophobic part is formed by one (or more) aliphatic chains $CH_3(CH_2)_n$. The hydrophilic part can be an ion (either anion or cation) which forms a "polar head". The polar head has an affinity to liquids with high dielectric constant such as water. Surfactants modifying the spreading of liquids on surfaces are of primary importance in various fields of industry and a lot of literature is devoted to them.[23] They also govern a diversity of phenomena related to the wetting of real surfaces, such as superspreading which will be discussed further.

1.9 A BIT OF EXOTICS: WHEN THE SURFACE TENSION IS NEGATIVE?

Consider the situation when a chemical reaction between two immiscible liquids creates surfactant molecules at the interface between them. In this case, the interfacial surface tension decreases with increasing amount of the surfactant.[24] The overpopulation of the interface by surfactants can give rise to a negative surface tension, when an interfacial reaction is faster than the time scale of system's equilibration. Other mechanisms which can make the interfacial tension to be transiently negative have been discussed in the context of micro-emulsions and spontaneous emulsification.[25,26]

In contrast to the positive surface tension, the negative γ tends to stretch the interface, giving rise to its roughening and bringing into existence a variety of interfacial structures ranging from ripples to micelle-like formations.[24] Remarkably, the first discussion of the exotic case of the negative surface tension took place 100 years ago.[27]

1.10 THE LAPLACE PRESSURE

Surface tension leads to the important and widespread phenomenon of overpressure existing in the interior of drops and bubbles.[17] Let us consider a drop of liquid 1 placed in liquid 2 (see Figure 1.7).

The drop is supposed to be in equilibrium. The minimal surface energy of a drop corresponds to its spherical shape of radius R. Assume that the pressure in the drop is p_1 and the pressure outside the drop is p_2. If the interface between liquids is displaced

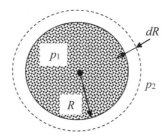

FIGURE 1.7 A droplet of liquid 1 of the radius R is in equilibrium with the surrounding liquid 2.

by an amount of dR (see Figure 1.9), according to the principle of virtual works, the total work $\delta\tilde{A} = 0$. The total work is given by

$$\delta\tilde{A} = p_1 dV_1 + p_2 dV_2 - \gamma dS_{int}, \qquad (1.14)$$

where γ is the surface tension at the interface between liquids and S_{int} is the area of the interface separating the liquids. Considering $dV_1 = -dV_2 = 4\pi R^2 dR$, $dS_{int} = 8\pi R dR$ immediately yields:

$$p_1 - p_2 = p_L = \frac{2\gamma}{R}, \qquad (1.15)$$

where p_L is the Laplace overpressure. The well-known simplified Laplace formula is recognized. It should be emphasized that the Laplace overpressure may by calculated with Eq. (1.15) only for the simplest spherical surfaces. For more complicated surfaces, the Laplace overpressure is calculated with Eq. (1.16):

$$p_1 - p_2 = \gamma\left(\frac{1}{R_1} + \frac{1}{R_2}\right), \qquad (1.16)$$

where R_1 and R_2 are the principal radii of curvature of the surface. When we have a drop surrounded by vapor, it obtains the form $p_{liq} - p_{vap} = p_L = \gamma(\frac{1}{R_1} + \frac{1}{R_2})$, where p_{liq} and p_{vap} are the pressures of a liquid and vapor, respectively.[28,29] The meaning of the main radii of curvature of the surface is illustrated with Figure 1.8 presenting a dumbbell-like body.

We look for R_1 and R_2 in a certain point of the surface enclosing the dumbbell and characterized by a normal vector $\delta\vec{\varsigma}$. For calculating R_1 and R_2 we have to cut our surface with two mutually orthogonal planes intersecting each other along $\delta\vec{\varsigma}$ (see Figure 1.8). The intersection of these planes with the interface defines two curves, the radii of curvature of which are R_1 and R_2. The radii of curvature could be positive or negative. R is defined as positive if the center of the corresponding circle lies inside the bulk, and negative if otherwise. The curvature of the surface $\hat{C} = 1/R_1 + 1/R_2$ is independent of the orientation of the planes. For a *spherical* droplet $R_1 = R_2 = R$ and consequently for the Laplace pressure jump, we have $p_1 - p_2 = p_{liq} - p_{vap} = 2\gamma/R$.

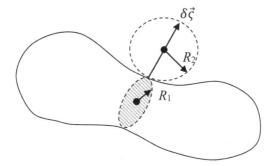

FIGURE 1.8 Scheme depicting the main radii of curvature of a dumbbell-like surface.

1.11 DIMENSIONLESS NUMBERS IMPORTANT FOR UNDERSTANDING BIOINSPIRED MATERIALS

Eq. (1.15) enables to introduce the dimensionless number quantifying interplay between surface phenomena and gravity. The maximal hydrostatic pressure (which is due to gravity) in a spherical droplet equals $p_h = 2\rho g R$; the Laplace pressure, which is due to the surface tension, in turn, equals $p_L = \frac{2\gamma}{R}$, let us equate these pressures and obtain the following equation introducing the so-called capillary length l_{ca}:

$$R = l_{ca} = \sqrt{\frac{\gamma}{\rho g}}. \qquad (1.17)$$

Quite remarkably, the value of l_{ca} is of the order of magnitude of a few mm for a majority of liquids even for mercury for which both ρ and γ are large. When $R \ll l_{ca}$, the effects due to gravity are negligible; in other words, a droplet will keep a shape close to spherical; when $R \cong l_{ca}$, the effects due to gravity become considerable. Eq. (1.17) may be re-written in the dimensionless form, introducing the dimensionless Bond number (also known as the Eötvös number) namely:

$$Bo = \frac{\rho g L^2}{\gamma}, \qquad (1.18)$$

where L is the characteristic length scale, which in the case of the droplet deposited on the solid substrate obviously equals the radius of the droplet R; hence, $Bo = \frac{\rho g R^2}{\gamma}$. When $Bo \ll 1$, the effects due to gravity are negligible and the shape of the droplet is dictated solely by the surface tension.

We already introduced two physical values crucially important for understanding of the physics of bioinspired materials, namely: viscosity and surface tension. The dimensionless value describing interrelation between viscous phenomena and surface-tension-inspired ones is called the capillary number and it is defined according to the following equation:

$$Ca = \frac{\eta v}{\gamma}, \qquad (1.19)$$

where v is the characteristic velocity and η is the viscosity of the liquid. When $Ca \ll 1$ takes place, the effects due to the viscosity are negligible, when compared to the surface tension governed ones.

1.12   SURFACE TENSION OF SOLIDS

Unlike the situation with liquids, the surface tension of solids is not necessarily equal to the surface free energy. We can imagine the process of forming a fresh surface of condensed phase divided into two steps: first, the material is cleaved, keeping the atoms fixed in the same positions that they occupied in the bulk; second, the atoms in the surface region are allowed to rearrange themselves to their final equilibrium positions. In the case of liquid, these two steps typically occur as once, due to the high mobility of liquid molecules, but with solids the second step may occur only slowly due to the low mobility of molecules constituting the surface region.[2,5] Thus, it is possible to stretch a surface of a solid without changing the number of atoms in it, but only their distances from one another.

Thus, the surface-stretching tension (or surface stress) τ is defined as the external force per unit length that must be applied to retain the atoms or molecules in their initial equilibrium positions (equivalent to the work spent in stretching the solid surface in a two-dimensional plane); whereas a *specific* surface free energy $\hat{G}_S$ is the work spent in forming a unit area of a solid surface. The relation between surface free energy and stretching tension could be derived as follows. For an anisotropic solid, if the area is increased in two directions by dS_1 and dS_2, the relation between τ_1, τ_2 and the free energy per unit area $\hat{G}_S$ is given by

$$\tau_1 = \hat{G}_s + S_1 \frac{d\hat{G}_s}{dS_1}; \tau_2 = \hat{G}_s + S_2 \frac{d\hat{G}_s}{dS_2}. \tag{1.20}$$

If the solid surface is isotropic, Eq. (1.20) reduces to

$$\tau = \frac{d\left(S\hat{G}_S\right)}{dS} = \hat{G}_S + S\frac{d\hat{G}_S}{dS}. \tag{1.21}$$

For liquids, the last term in Eq. (1.21) is zero; hence, $\tau = \hat{G}_S = \gamma$.

1.13   VALUES OF SURFACE TENSIONS OF SOLIDS

De Gennes et al. proposed to divide all solid surfaces into two large groups (see Reference 29).

1. High-energy surfaces. These are surfaces possessing the surface energy $\hat{G}_S \approx 200 - 5,000\,\text{mJ/m}^2$. High-energy surfaces are inherent for materials built with strong chemical bonds such as ionic, metallic or covalent. For a covalent bond-built diamond, the surface energy could be approximately equaled to the half of the energy required to break the total number of

covalent bonds passing through a unit of cross-sectional area of the material.[3] The appropriate calculation supplies the value of $5,670\,mJ/m^2$. For ionic and metallic solids, the calculations are more complicated; for the values of surface energies of various solids, see Reference 16.

2. Low-energy solid surfaces. These are surfaces possessing the surface energy $10-50\,mJ/m^2$. Low-energy solid surfaces are inherent for solids based on the relatively weak van der Waals chemical bonds, such as in polymers. As it was already shown in Section 1.5, the London dispersion force dominates in the van der Waals forces. Thus, the estimation $\hat{G}_s \approx const/d_m^2$ will be valid for solids built on the van der Waals forces. Moreover, a straightforward calculation of the energy of the London interaction given by Eq. (1.9) supplies the value of k_BT (Reference 16). Hence, for a rough estimation of the surface energy of this kind of solids, we can take $\hat{G}_s \approx k_BT/d_m^2$. This simple formula explains the surprising proximity of specific surface energies of very different solids and liquids, such as plastics and organic solvents. For example, the specific surface energy of polystyrene equals $32-33\,mJ/m^2$ (compare this value with surface tensions of organic solvents supplied in Table 1.1).[16]

1.14 SURFACE TENSION AND FLOATING

Physics of floating, governed to a much extent by the surface tension (like many other fundamental physical phenomena), was first studied by the great Greek philosopher Archimedes. Today, the thought process of this remarkable person is clearer to us since the discovery of the "Archimedes Palimpsest" (recycled parchment), which contains 10th-century Greek versions of seven texts by Archimedes, which were copied by an unknown writer using iron gall ink. This manuscript, which was later overwritten by a Greek liturgical text, is the most ancient source of several treatises by Archimedes. It was deemed lost until 1998, when it was sold to a private collector who then entrusted a museum with its restoration and study. For more details of the detective story surrounding the Archimedes Palimpsest, see References 30 and 31. In his treatise "On Floating Bodies", one of the most important texts of the Palimpsest, the famous Archimedes principle is formulated as follows: "Any body wholly or partially immersed in a fluid experiences an upward force (buoyancy) equal to, but opposite to the weight of the fluid displaced". It follows from this principle that an object can float only if it is less dense than the liquid in which it is placed. However, a sure-handed child may place a steel needle on a water surface and it will float, as shown in Figure 1.2. The first explanation for this effect is credited to Galileo Galilei.[32,33]

Consider a heavy plate placed on a water surface, as shown in Figure 1.9.

When the three-phase line (appearing when solid, liquid and gaseous phases contact) is firmly pinned to the surface of the plate, it may displace a volume which is much larger than the total volume of the plate itself, as shown in Figure 1.9A. Hence, the buoyancy will be essentially increased. It is seen, however, that Galilei related the floating of heavy bodies to the increase of the Archimedes force only. This explanation is at least partially true. The unbelievable scientific intuition of Galilei is admirable, but actually the floating of heavy objects arises through the interplay of buoyancy and surface tension. The restoring force that counteracts the floating body weight

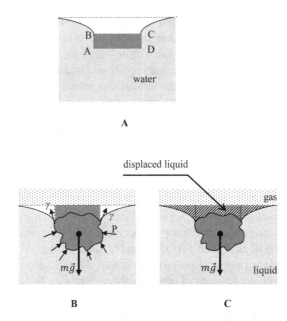

FIGURE 1.9 (**A**) Galileo intuitive reasoning explaining the floating of the heavy plate ABCD. When the triple line is pinned to the surface of the plate, it may displace the volume which is much larger than the total volume of the plate itself. (**B**) The buoyancy of the floating heavy object equals to the total weight of the displaced water, shown shaded, dark-gray. The surface tension force which equals γ for the unit length of the triple line also supports floating. (**C**) The total restoring force, including the surface tension and buoyancy, equals the total weight of the displaced liquid (the gray shaded area).

$m\vec{g}$ comprises the surface-tension-inspired force and the buoyancy force arising from hydrostatic pressure, as shown in Figure 1.9B. Following Galilei idea's (illustrated in Figure 1.9A), the buoyancy equals the total weight of the liquid displaced by the body, shown with the dark-gray shading in Figure 1.9C. Hand-in-hand with the buoyancy, the surface tension also supports floating as shown in Figure 1.9B.

The American mathematician Joseph Bishop Keller demonstrated that the total force supporting the floating is equal to the weight of the entire volume of liquid displaced by the body, as depicted in Figure 1.9C.[34] By a remarkable coincidence, this fascinating theorem was published in 1998, the same year the Archimedes Palimpsest was re-discovered. Profound mathematical treatment of the surface-tension-supported floating is found in Reference 33. Understanding of this phenomenon is important for understanding of the locomotion of water striders, inspiring development of novel biomimetic materials. Propulsion of water striders will be addressed below.

1.15 VISCOUS DRAG

When a body moves within a fluid, a force is acting opposite to the relative motion of a body. This force is called the viscous drag force (the detailed discussion of the phenomenon of drag will be carried out in Section 5.1 in its relation to the "shark skin effect"). The drag force depends strongly on the value of the Reynolds number which is defined as follows:

$$Re = \frac{\rho v D}{\eta}, \tag{1.22}$$

where ρ and η are the density and viscosity of the liquid correspondingly; v and D are characteristic velocity and linear dimension of the body, respectively. We already acquainted the Bond and capillary dimensionless numbers in Section 1.11; now let us acquaint the Reynolds number. The Reynolds number is the ratio of inertial forces to viscous forces, as it is easily seen from Eq. (1.22) (see Reference 9). Eq. (1.22) may be re-written as follows:

$$Re = \frac{vD}{\tilde{v}_{kin}}, \tag{1.23}$$

where $\tilde{v}_{kin}$ is the kinematic viscosity of the liquid defined as $\tilde{v}_{kin} = \frac{\eta}{\rho}; [\tilde{v}_{kin}] = \frac{m^2}{s}$.

APPENDIX 1.A THE SHORT-RANGE NATURE OF INTERMOLECULAR FORCES

This appendix is important for understanding the entire materials science and not only the surface tension (energy) of condensed matter. The Keesom, Debye and London dispersion forces introduced in Section 1.5 all decrease with the distance as $\approx 1/r^6$. All these forces contribute to the so-called *van der Waals forces* acting between molecules. The power law index -6 is of a primary importance for constituting bulk and surface properties of condensed phases. Due to this power law, the total interaction of the molecule with other molecules is defined by neighboring ones and the contribution of the distant molecules is negligible. Let us discuss a cubic vessel L containing molecules with a diameter d_m attracting through a potential $U(r) = -\tilde{C}/r^n$ where $\tilde{C}$ is the constant and n is an integer. Let us also suppose that the number density of molecules $\tilde{\rho}$ is constant. Let us estimate the total energy of interaction of *one particular molecule* with all the other molecules in the vessel denoted as U_{int}^{total}.

$$U_{int}^{total} = \int_{d_m}^{L} U(r)\tilde{\rho}4\pi r^2 dr = -4\pi\tilde{C}\tilde{\rho}\int_{d_m}^{L} r^{2-n} dr = -\frac{4\pi\tilde{C}\tilde{\rho}}{(n-3)d_m^{n-3}}\left[1-\left(\frac{d_m}{L}\right)^{n-3}\right]. \tag{1.24}$$

Taking into account $\frac{d_m}{L} < 1$, we recognize that long-range contributions from distant molecules will disappear only for $n>3$. When $\frac{d_m}{L} < 1$, $n > 3$, we obtain:

$$U_{int}^{total} = -\frac{4\pi\tilde{C}\tilde{\rho}}{(n-3)d_m^{n-3}}. \tag{1.25}$$

But for $n < 3$, we have $\left(\frac{d_m}{L}\right)^{n-3}$ greater than unity, and for $L \gg d_m$, the contribution from distant molecules will dominate over neighbor ones (for $n = 3$, Formula (1.24) gives $U_{int}^{total} \sim log\,(d_m/L)$, which is usually considered as long-ranged). When $n > 3$, the size of the system should not be taken into account and some of the thermodynamic properties such as pressure and temperature turn out to be *intensive*. Thus, we see that the power index $n = 6$ turns out to be of primary importance allowing us to neglect distant interactions between molecules. However, we will see later that in certain cases, the range of intermolecular forces between liquid layers can extend out to 100 nm.

APPENDIX 1.B ACCURATE THERMODYNAMIC DEFINITION OF A SURFACE TENSION

When the surface of a liquid is changed under constant temperature, the work necessary for this change is given by $d\tilde{A} = -dF_S$, where F_S is a free-surface Helmholtz energy (defined as $F_S = W_S - TS_S$, where W_S and S_S are the surface energy and entropy, respectively; see Section 1.7). Thus, we have for a surface energy:

$$W_S = F_S + TS_S = \left(\gamma - T\frac{d\gamma}{dT}\right) \times (surface\ area). \qquad (1.26)$$

It is easily seen from Eq. (1.26) that the surface tension γ could not be interpreted as an energy per unit area of a surface (as if often mistakenly accepted in literature), but it is a *free Helmholtz energy per unit area of a surface*.

Bullets
- Surface tension is a tension in a surface due to the unusual energetic state of the surface molecules.
- For liquids at constant temperature and pressure and in equilibrium, the surface tension is physically equivalent to the specific surface Helmholtz free energy.
- The surface tension of solids is not necessarily equal to the surface free energy.
- Surface tension is stipulated mainly by the London dispersion forces and metallic or hydrogen bonds (when they are present).
- The surface tension of most organic and non-organic liquids at room temperature is within $20 - 70\,\text{mJ/m}^2$.
- Surface tension is temperature-dependent. The temperature slope of the surface tension for simple liquids is negative. The temperature slope of the surface tension for simple liquids may be positive due to the entropy component of the free energy. This occurs when liquid at the surface is more ordered than in the bulk. The effect is called "the surface freezing".
- Surface tension leads to the Laplace overpressure existing in the interior of drops and bubbles. $p_L = p_1 - p_2 = \gamma\left(\frac{1}{R_1} + \frac{1}{R_2}\right)$.
- Viscosity is a measure of a fluid's resistance to flow. It describes the internal friction of a moving fluid.
- Viscosity may be seen as the physical property which characterizes momentum transport within a fluid or gas.

- Viscosity of liquids is exponentially temperature-dependent.
- The Arrhenius-like exponential temperature dependence of viscosity is due to the activation nature of flow: displacement of a molecule of liquid demands surmounting of a potential barrier, when a molecule of liquid jumps from one equilibrium position to another.

EXERCISES

1. Two droplets of mercury with the radius $r = 1$ mm coalesced in one big droplet. Calculate the heat released under the coalescence.

   Solution: The heat Q will be released under coalescence mainly due to the change in the surface energy. Neglecting contributions of other kinds of energy (for example, the possible change in the gravitational energy) related to the coalescence of droplets, we have $Q = \gamma S_1 - \gamma S_2 = 4\pi\gamma(2r^2 - R^2)$, where S_1 and S_2 are the surface areas of small and big droplets, R is the radius of a big droplet and $\gamma_{merc} = 486$ mJ/m^2 is the surface tension of mercury. Volume conservation yields $2\frac{4}{3}\pi r^3 = \frac{4}{3}\pi r^3 \Rightarrow R = \sqrt[3]{2}r$. Finally for the heat released under the coalescence, we have $Q = 8\pi\gamma r^2\left(1 - 2^{-\frac{1}{3}}\right) = 2.5\mu$J.

2. Rod with the length of $L=5$ cm is free to slip along the rectangular frame, as shown in the picture. The opaque field in the picture is filled by thin liquid film with the surface tension $\gamma = 70$mJ/m^2; $h = 5$ cm (Figure 1.10).

   What mass m will keep the system in the equilibrium? What is the energy of the system G related to the surface tension?

   Answer: $m = \frac{2L\gamma}{g} = 0.07$ g; $G = 2\gamma Lh = 0.35$ mJ.

3. The soap bubble is blown. Blowing starts from the diameter $d_1 = 1$ cm and finishes at $d_2 = 11$ cm. What work should be done for blowing? The surface tension of the soap water is $\gamma = 50.0$ mJ/m^2.

   Solution: The work A may be estimated as $A = 2\pi\gamma(d_2^2 - d_1^2)$.

4. Basing of the considerations of dimensions estimates the period of oscillations T of a small droplet r possessing the density of ρ and surface tension of γ.

   Answer: $T \sim \sqrt{\frac{\rho r^3}{\gamma}}$.

5. Calculate the curvature of the surfaces supplied in Figures 1.11A,B. Surface A is the cylindrical one; surface B is the saddle.

   Answer: (A) $\hat{C} = \frac{1}{R}$; (B) $\hat{C} = \frac{1}{R_1} - \frac{1}{R_2}$.

6. Consider the Carnot engine, exploiting a liquid film as a working medium. Calculate the temperature derivative of the surface tension.

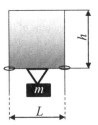

FIGURE 1.10   Exercise 2. Rod slipping along the rectangular frame and forming the liquid film is shown.

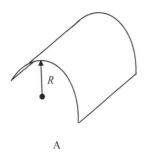 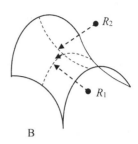

A   B

FIGURE 1.11 Exercise 5.

Hints: (1) Suppose the temperatures of "warmer" and "cooler" of the Carnot machine are close. (2) The heat dQ necessary for the isothermal stretching of the film is $dQ = \tilde{q}dS_{int}$, where S_{int} is an area of a liquid film and $\tilde{q}$ is the specific heat necessary for isothermal stretching of the unit area of the liquid film.

Answer: The analysis of the Carnot cycle yields $\frac{d\gamma}{dT} = -\frac{\tilde{q}}{T}$.

7. Demonstrate that in the vicinity of 0 K, γ becomes independent of the temperature, namely: $\lim\limits_{T \to 0} \frac{d\gamma}{dT} = 0$ takes place.

   Hints: Involve the equation $\frac{d\gamma}{dT} = -\frac{\tilde{q}}{T}$ obtained in the previous problem and involve $\tilde{q} = T\Delta\tilde{S}$, where $\Delta\tilde{S}$ is the entropy change, corresponding to increasing the surface by the unit area. Exploit the third law of thermodynamics.

8. Student shakes the pellet of mercury in its thermometer. He shakes the thermometer with the acceleration of $a \cong 10g$. The height of the mercury pellet $h = 5$ cm; $\gamma_{merc} = 486\, mJ/m^2$ and $\rho_{merc} = 13.6 \times 10^3\, kg/m^3$ are the surface tension and the density of mercury, respectively. Estimate the radius of the capillary tube r.

   Hint: Equal the inertia force acting on the mercury pellet to the surface-tension-related force, namely $\gamma 2\pi r$.

   Answer: $r \cong \frac{\gamma_{merc}}{5\rho_{merc}gh} \cong 1.5 \times 10^{-3}$ cm.

9. Explain the qualitatively the thermal dependence of the surface tension.
10. Explain the difference in the surface tension of liquids and solids.
11. Demonstrate that the surface tension of liquid may be approximately expressed as $\cong \frac{1}{6}\rho d_m \Delta H_{mass}$, where ΔH_{mass} is the specific mass enthalpy of vaporization (evaporation enthalpy per unit of mass); d_m is the diameter of the molecule.
12. What is the physical meaning of viscosity.
13. Explain the difference in the thermal dependence of viscosity and surface tension.
14. The field of velocities within the volume of liquid V is given by $\vec{u}(x,y,z)$. The rate of the energy dissipation $\dot{E}; [\dot{E}] = \frac{J}{s}$ within volume V is given by

$$\dot{E} = \eta \int\limits_{V} (\nabla\vec{u})^2\, dV.$$

Check this formula from the point of view of dimensions.

15. Explain the physical meaning of the Bond, Capillary and Reynolds numbers.
16. Calculate the capillary length for water and mercury.

REFERENCES

1. Bhushan B. Biomimetics, in Bioinspired Hierarchical-Structured Surfaces for Green Science and Technology, 3rd Edition, *Springer Nature*, Cham, Switzerland, 2018.
2. Adamson A. W., Gast A. P. *Physical Chemistry of Surfaces*, Sixth Edition, Wiley-Interscience Publishers, New York, 1990.
3. Erbil H. Y. *Surface Chemistry of Solid and Liquid Interfaces*, Blackwell, Oxford, 2006.
4. Marmur A., Ras R. H. A. *Non-Wettable Surfaces: Theory, Preparation and Applications* (Soft Matter Series, Volume 5), RSC, Cambridge, UK, 2017.
5. Kim H. J. *Biomimetic Microengineering*, CRC Press, Boca Raton, USA, 2020.
6. Guo H. Y., Li Q., Zhao H. P., Zhou K., Feng X. Q. Functional map of biological and biomimetic materials with hierarchical surface structures. *RSC Adv.* 2015, **5**, 66901–66926.
7. Sun T., Feng L., Gao X., Jiang L. Bioinspired surfaces with special wettability. *Acc. Chem. Res.* 2005, **38** (8), 644–652.
8. Nosonovsky M., Bhushan B. Superhydrophobic surfaces and emerging applications: Non-adhesion, energy, green engineering. *Current Opin. Colloid Interface Sci.* 2009, **14** (4), 270–280.
9. Bormashenko E. *Physics of Wetting. Phenomena and Applications of Fluids on Surfaces*, De Gruyter, Berlin, 2017.
10. Moore J. C., Kazachkov A., Anders A. G., Willis C. The Danger of Misrepresentations in Science Education, in Informal Learning and Public Understanding of Physics, 3rd International GIREP Seminar 2005. Selected Contributions; Planinsic G., Mohoric A., Eds.; University of Ljubljana, Ljubljana, Slovenia, 2005; pp. 399–404.
11. Hirai N., Eyring H. Bulk viscosity of liquids. *J. Appl. Phys.* 1958, **29**, 810.
12. Gutmann F., Simmons L. M. The temperature dependence of the viscosity of liquids. *J. Appl. Phys.* 1952, **23**, 977.
13. Messaâdi A., Dhouibi N., Hamda H., Belgacem F. B. M., Adbelkader Y. H., Ouerfelli N., Hamzaoui A. H. A new equation relating the viscosity Arrhenius temperature and the activation energy for some Newtonian classical solvents. *J. Chem.* 2015, **2015**, 163262.
14. Mauro J. C., Yue Y., Ellison A. J., Gupta P. K., Allan D. C. Viscosity of glass-forming liquids. *PNAS* 2009, **106** (47), 19780–19784.
15. Ikeda M., Aniya M. Bond strength - coordination number fluctuation model of viscosity: An alternative model for the Vogel-Fulcher-Tammann Equation and an application to bulk metallic glass forming liquids. *Materials* 2010, **3**, 5246–5262.
16. Israelachvili J. N. *Intermolecular and Surface Forces*, Third Edition, Elsevier, Amsterdam, 2011.
17. Bormashenko E. Why are the values of the surface tension of most organic liquids similar? *Am. J. Phys.* 2010, **78**, 1309–1311.
18. Su Y. Z., Flumerfelt R. W. A continuum approach to microscopic surface tension for the n-alkanes. *Ind. Eng. Chem. Res.* 1996, **35**, 3399–3402.
19. Fowkes F. M. Additivity of intermolecular forces at interfaces. *J. Phys. Chem.* 1962, **67**, 2538–2541.
20. Van Oss C. J., Good R. J., Chaudhury M. K. Additive and nonadditive surface tension components and the interpretation of contact angles. *Langmuir* 1988, **4**, 884–891.
21. Ocko B. M., Wu X. Z., Sirota E. B., Sinha S. K., Gang O., Deutsch M. Surface freezing in chain molecules: Normal alkanes. *Phys. Rev. E* 1997, **55**, 3164–3181.
22. Sloutskin E., Bain C. D., Ocko B. M., Deutsch M. Surface freezing of chain molecules at the liquid–liquid and liquid–air interfaces. *Faraday Discuss.* 2005, **129**, 339–352.
23. Schramm L. L. *Emulsions, Foams and Suspensions, Fundamentals and Applications*, Wiley, Weinheim, 2005.

24. Patashinski A. Z., Orlik R., Paclawski K., Ratner M. A., Grzybowski B. A. The unstable and expanding interface between reacting liquids: Theoretical interpretation of negative surface tension. *Soft Matter* 2012, **8**, 1601–1608.
25. Granek R., Ball R. C., Cates M. E. Dynamics of spontaneous emulsification. *J. Phys. II France* 1993, **3** (6), 829–849.
26. Guttman S., Sapir Z., Schultz M., Butenko A. V., Ocko B. M., Deutsch M., Sloutskin E. How faceted liquid droplets grow tails. *PNAS* 2016, **113**, 493–496.
27. Kimball A. L. Negative surface tension. *Science* 1917, **45**, 85–87.
28. Landau L., Lifshitz E. *Fluid Mechanics*, 2nd Edition, Butterworth-Heinemann, Oxford, UK, 1987.
29. de Gennes P. G., Brochard-Wyart F., Quéré D. *Capillarity and Wetting Phenomena*, Springer, Berlin, 2003.
30. Netz R., Noel W. *The Archimedes Codex: How a Medieval Prayer Book Is Revealing the True Genius of Antiquity's Greatest Scientist*, Da Capo Press, New York, 2007.
31. Salerno E., Tonazzini A., Bedin L. Digital image analysis to enhance underwritten text in the Archimedes palimpsest. *Int. J. Doc. Anal. Recognit. (IJDAR)* 2007, **9**, 79–87.
32. Galilei G., *Discourse concerning the natation of bodies*, 1663. Translated into English by Thomas Salusbury, University of Illinois Press, Urbana, Illinois, USA, 1960.
33. Vella D. Floating versus sinking. *Annu. Rev. Fluid Mech.* 2015, **47**, 115–135.
34. Keller J. B. Surface tension force on a partially immersed body, *Phys. Fluids*, 1998, **10**, 3009–3010.

2 Wetting of Real Surfaces—Lotus Effect

In this chapter, we will speak about bioinspired (or biomimetic) superhydrophobic and omniphobic materials developed in last decades. We will discuss the Lotus effect and superhydrophobic materials emerging from the Lotus effect, which are already available in the industry. It should be emphasized that superhydrophobic material which is already available and broadly used in the industry emerged from the synthesis of approaches developed recently by biomimetic and nano-science. First of all, we will discuss the scientific foundations of wetting of real surfaces, which are essential for understanding the Lotus effect. We will unravel the physical mechanisms underlying the Lotus effect and survey biomimetic materials exploiting these effects and their industrial applications. Technological demands to the superhydrophobic materials will be defined.

2.1 WETTING OF SURFACES: THE CONTACT ANGLE—WHAT IS WETTING? THE SPREADING PARAMETER

The effect of *superhydrophobicity* was first observed in nature on *lotus leaf* and some other plant leaves that would not get wet. This discovery gave rise to a diversity of novel, non-wettable and self-cleaning materials, coatings and paints. In order to understand the phenomenon of "superhydrophobicity", we have to acquaint first the physical groundings of wetting.

We already introduced in Chapter 1 the notion of the surface tension and discussed the behavior of "free droplets" which do not contact surfaces. Now, we address wetting of surfaces. Wetting is the ability of a liquid to maintain contact with a solid surface, resulting from intermolecular interactions when the two are brought together. The idea that the wetting of solids depends on the interaction between particles constituting a solid substrate and liquid has been expressed explicitly in the famous essay by Thomas Young.[1] When a liquid drop is placed on the solid substrate, two main *static* scenarios are possible: either liquid spreads completely or it sticks to the surface and forms a cap as shown in Figure 2.1A (a solid surface may be flat or rough, homogenous or heterogeneous).

The precise definition of the contact angle will be given later; at this stage, we only require that the radius of the droplet should be much larger than the characteristic scale of the surface roughness. The observed wetting scenario is dictated by the spreading parameter

$$\Psi = \hat{G}_{SA}^* - \left(\hat{G}_{SL}^* + \hat{G}_{LA} \right), \tag{2.1}$$

where $\hat{G}_{SA}^*$ and $\hat{G}_{SL}^*$ are the *specific* surface energies at the rough solid/air and solid/liquid interfaces, respectively (the asterisk reminds us that $\hat{G}_{SA}^*$ and $\hat{G}_{SL}^*$ do not coincide

DOI: 10.1201/9781003178477-2

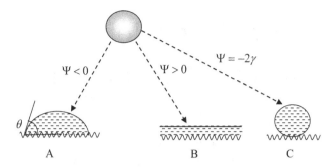

FIGURE 2.1 The three wetting scenarios for sessile drops: partial wetting (**A**), complete wetting (**B**) and complete dewetting (**C**).

with the specific surface energies of smooth surfaces $\hat{G}_{SA}$, $\hat{G}_{SL}$ introduced in Section 1.12), and $\hat{G}_{LA} = \gamma$ is the specific energy of the liquid/air interface, coinciding with the surface tension as demonstrated in Section 1.12. The dimension of the spreading parameter is $[\Psi] = \frac{J}{m^2}$. When $\Psi > 0$, total wetting is observed, depicted in Figure 2.1B. The liquid spreads completely in order to lower its surface energy ($\theta = 0$). When $\Psi < 0$, the droplet does not spread but forms a cap resting on a substrate with a contact angle θ, as shown in Figure 2.1A. This case is called *partial wetting*. When the liquid is water, surfaces demonstrating $\theta < \frac{\pi}{2}$ are called *hydrophilic*, while surfaces characterized by $\theta > \frac{\pi}{2}$ are referred as *hydrophobic*. One more extreme situation is possible, when $cos\theta = -1$, such as depicted in Figure 2.1C. This is the situation of *complete dewetting* or *superhydrophobicity* or "*lotus effect*", which gave rise to numerous novel materials, to be discussed below in much detail. When the solid surface is atomically flat, chemically homogeneous, isotropic, insoluble, non-reactive and non-stretched (thus, there is no difference between the specific surface energy and surface tension, as explained in Section 1.12), the spreading parameter obtains its convenient form:

$$\Psi = \gamma_{SA} - (\gamma_{SL} + \gamma),\qquad(2.2)$$

where γ_{SA}, γ_{SL} and γ are the surface tensions at the solid/air (vapor), solid/liquid and liquid/air interfaces, respectively.[2] When the droplet forms a cap, the line at which solid, liquid and gaseous phases meet is called the *triple* or (three-phase) *line*.

2.2   THE YOUNG EQUATION

We will start from wetting of an *ideal* surface, i.e., an atomically flat, chemically homogeneous, isotropic, insoluble, non-reactive and non-deformed solid surface in the situation where $\Psi < 0$. The contact angle of such a surface (also called the Young angle and denoted below as θ_Y) is given by the famous Young equation:

$$cos\theta_Y = \frac{\gamma_{SA} - \gamma_{SL}}{\gamma}.\qquad(2.3)$$

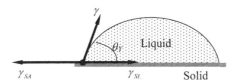

FIGURE 2.2   Scheme illustrating the Young equation.

It is noteworthy that a thoughtful reader will not reveal this equation in the paper by Thomas Young (Reference 1). Actually, it was formulated much later by Dupre. The Young equation may be interpreted as the balance of capillary forces acting on the unit length of the *triple line*, as shown in Figure 2.2. Indeed, projecting these forces on the horizontal plane immediately yields:

$$\gamma \cos \theta_Y = \gamma_{SA} - \gamma_{SL}. \tag{2.4}$$

The "force-based" interpretation of the Young equation is simple, however, problematic. An accurate grounding of the Young equation is attainable within the variational analysis of the thermodynamics of wetting, discussed in detail in Reference 3. Comparing Eq. (2.4) with Eq. (2.2) supplies the useful formula:

$$\Psi = \gamma (cos\theta_Y - 1). \tag{2.5}$$

It could be recognized that in the situation of *complete dewetting* or *superhydrophobicity ("lotus effect")*, shown in Figure 2.1C, $\Psi = -2\gamma$. This result is intuitively clear: indeed, in the situation of complete dewetting, there is no actual contact of a droplet with a solid surface and the spreading parameter is totally defined by the liquid/air surface specific energy. Actually, this situation on *flat* surfaces is unachievable, but it exists on *rough* surfaces, as it will be shown in Section 2.12.

Take a close look at Eq. (2.3). It asserts that the contact angle is unambiguously defined by the triad of surface tensions: γ_{SA}, γ_{SL} and γ, as it was stated first by Sir Thomas Young: "For each combination of a solid and a fluid, there is an appropriate angle of contact between the surfaces of the fluid, exposed to the air, and to the solid".[1] The *Young contact angle* θ_Y is the *equilibrium* contact angle that a liquid makes with an *ideal* solid surface.[3]

Remarkably, the Young contact angle is independent of the shape of a droplet and also on external fields (under very general assumptions about the nature of these fields), including gravity. This counter-intuitive statement may be proved by the solution of the variational problem of wetting with free endpoints.[4]

Notice that for droplets or surfaces with very small radii of curvature deposited on the *ideal* surfaces, the equilibrium contact angle may be different due to line tension.

2.3   LINE TENSION

Surface tension is due to the special energy state of the molecules at a solid or liquid surface and it is important for understanding of the Lotus effect. Molecules located at the triple (three-phase) line where solid, liquid and gaseous phases meet are also in an unusual energy state. The notion of line tension has been introduced by Gibbs.

Gibbs stated: "These (triple) lines might be treated in a manner entirely analogous to that in which we have treated surfaces of discontinuity. We might recognize linear densities of energy, of entropy, and of several substances which occur about the line, also a certain linear tension".[5,6] In spite of the fact that the concept of line tension is intuitively clear, it remains one of the most obscure and disputable notions in surface science. It was suggested (see Reference 7) that the line tension Γ (the dimension of the line tension $\lfloor\Gamma\rfloor=\frac{J}{m}=N$) may be split into "interactional" (denoted below Γ_{int}) and entropy-inspired (denoted below Γ_{en} contributions):

$$\Gamma = \Gamma_{int} + \Gamma_{en}. \qquad (2.6)$$

A very simple estimation of the entropy contribution to the line tension looks as follows:

$$\Gamma_{en} \cong \frac{k_B T}{d_m}, \qquad (2.7)$$

where k_B is the Boltzmann constant and d_m is the molecular dimension.[7,8] More accurate estimation of the entropy contribution to the line tension arises from the assimilation of water molecules to the polymer chain and is formulated as follows:

$$\Gamma_{en} \cong \frac{k_B T}{\xi}, \qquad (2.8)$$

where $\xi > d_m$ is the correlation length of the polymer chain.[7,9] Eqs. (2.7) and (2.8) hint to the "entropy origin" of the linear tension.[9] Indeed, the so-called "entropic forces" grow linearly with temperature.[9] The notion of the line tension remains one of the most controversial in the interface science. The researchers disagree not only about the value of the line tension but even about its sign. Estimation of the line tension arising from Eq. (2.8) yields $\Gamma \cong 1.5\times10^{-11}$ N at ambient conditions. Experimental values of a line tension in the range of $10^{-5}-10^{-11}$ N were reported.[5] Very few methods allowing experimental measurement of line tension have been developed.[10] A. Marmur estimated a line tension as $\Gamma \cong 4d_m\sqrt{\gamma_{SA}\gamma}\cot\theta_Y$, where d_m is the molecular dimension, γ_{SA}, γ are the surface energies of the solid and liquid correspondingly and θ_Y is the Young angle. Marmur concluded that the magnitude of the line tension is less than 5×10^{-9} N, and that it is positive for acute and negative for obtuse Young angles.[11] However, researchers reported negative values of the line tension for hydrophilic surfaces.[10] As to the magnitude of the line tension, the values in the range $10^{-9}-10^{-12}$ N appear as realistic. Large values of the line tension Γ reported in the literature are most likely due to contaminations of the solid surfaces.[2]

Let us estimate the characteristic length scale l at which the effect of line tension becomes important by equating surface and "line" energies: $l \cong \frac{\Gamma}{\gamma} \cong 1-100$ nm. It is clear that the effects related to line tension can be important for nano-scaled droplets or for nano-scaled rough surfaces. When we develop "lotus-like", biomimetic, self-cleaning surfaces, they include micro- and nano-scaled elements of topography; thus, the effects due to the line tension may be essential. Behavior of nano-scaled droplets is of a primary importance for the development of anti-icing surfaces.[12]

The influence of the line tension on the contact angle of an axisymmetric droplet is considered in the so-called Boruvka-Neumann equation:

$$cos\theta = \frac{\gamma_{SA} - \gamma_{SL}}{\gamma} - \frac{\Gamma}{\gamma a},$$
(2.9)

where a is the contact radius of the droplet. Let us take a close look at Eq. (2.9); the Boruvka-Neumann equation predicts the dependence of the equilibrium contact angle on the contact radius of the droplet a. In other words, the equilibrium contact angle becomes dependent on the size of droplet and this is in striking contrast to the Young equation summarized by Eq. (2.3), in which the equilibrium contact angle does not depend on the droplet size (volume). Again, the equilibrium contact angle, predicted by Eq. (2.9) is independent of external fields, such as gravity.[4]

2.4 DISJOINING PRESSURE

One more notion is crucial for understanding wetting phenomena and this is the disjoining pressure. Now we address very thin liquid films deposited on ideal solid surfaces. If we place a film of thickness e (see Figure 2.3) on an ideal solid substrate, its specific surface energy will be $\gamma_{SL} + \gamma$.

However, if the thickness e tends to zero ($e \rightarrow 0$), we return to a bare solid with a specific surface energy of γ_{SA} (see Reference 2).

It is reasonable to present the specific surface energy of the film $\hat{G} = \frac{G}{S}$ (S is area, $[\hat{G}] = \frac{J}{m^2}$) as follows:

$$\hat{G}(e) = \gamma_{SL} + \gamma + \Omega(e),$$
(2.10)

where $\Omega(e)$ is a function of the film defined in such a way that $\lim_{e \rightarrow \infty} \Omega(e) = 0$ and $\lim_{e \rightarrow 0} \Omega(e) = \Psi = \gamma_{SA} - \gamma_{SL} - \gamma$ (see Eq. (2.2) and Reference 2). It could be demonstrated that when the molecules of solid and liquid interact via the van der Waals interaction (see Section 1.5 and References 13 and 14), $\Omega(e)$ obtains the form:

$$\Omega(e) = \frac{A}{12\pi e^2},$$
(2.11)

where A is the so called Hamaker constant, which is in the range of $A \cong 10^{-19} \div 10^{-20}$ J (see References 2, 13 and 14). Reference 13 supplies the extended discussion of the physical nature and values of the Hamaker constant. The Hamaker constant could be expressed as follows:

$$A = \pi^2 \varpi \tilde{\alpha}_L (\tilde{\alpha}_S - \tilde{\alpha}_L),$$
(2.12)

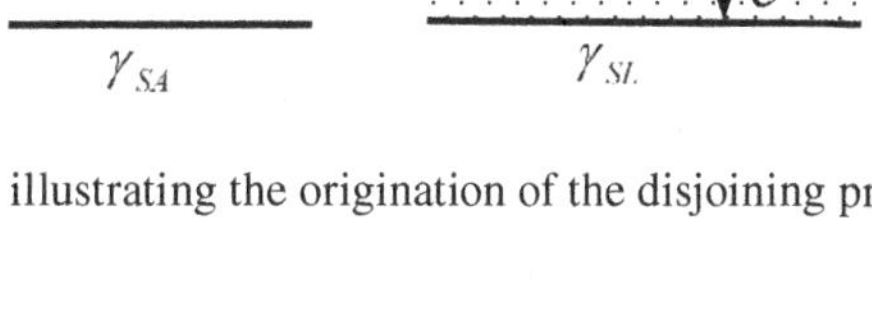

FIGURE 2.3 Scheme illustrating the origination of the disjoining pressure.

where $\tilde{\alpha}_L$ and $\tilde{\alpha}_S$ are the specific volume polarizabilities of liquid and solid substrates, respectively; ϖ is a constant that depends very little on the nature of solid and liquid.[2]

It could be seen from Eq. (2.12) that the Hamaker constant could be positive or negative. It will be positive when the solid has higher polarizability than the liquid $(\tilde{\alpha}_S > \tilde{\alpha}_L)$. This situation can happen on high-energy surfaces (see Section 1.13); the opposite case occurs on low-energy surfaces $(\tilde{\alpha}_S < \tilde{\alpha}_L)$. It could be seen from Eq. (2.10) that when $\Omega(e) < 0$ takes place, it diminishes the specific surface energy of the solid/thin liquid film system; thus, the van der Waals interaction will thin the film trying to cover as large a surface of the substrate as possible.

The negative derivative of $\Omega(e)$ is called the *disjoining pressure*:

$$\Pi(e) = -\frac{d\Omega}{de} = -\frac{A}{6\pi e^3},\qquad(2.13)$$

introduced into surface science by B. V. Derjaguin.[15,16] The disjoining pressure given by Eq. (2.13) is mainly due the London dispersion forces introduced in Section 1.5. Actually, in the framework of the continuum approach, the van der Waals (London dispersion) forces manifest themselves as external normal stresses ("disjoining pressures") imposed on interfaces.[17,18] The disjoining pressure plays a primary role in the theory of thin liquid films deposited on solid surfaces[17,18] and other wetting phenomena, including constitution of the apparent contact angle to be discussed in detail in Section 2.8.[19] For our discussion, it important that the disjoining pressure influences the stability of the "lotus effect" as demonstrated in References 20 and 21 and as it will be discussed below in detail.

Disjoining pressure remains one of the most obscure and intuitively unclear notions of the interface science; however, it is crucial for understanding of a diversity of wetting phenomena. It may be understood from the following qualitative considerations: consider the liquid film with the thickness of h_1 which is comparable with the characteristic range of intermolecular forces R_m, depicted in Figure 2.4. In other words, R_m is the distance, at which intermolecular forces become negligible. Consider the molecule labeled A located in the middle of the film, as depicted in Figure 2.4.

When $h_1 > 2R_m$ takes place, the energy of the molecule A equals the energy of any bulk water molecule. When the thickness of the film equals h_2 (see Figure 2.4), the

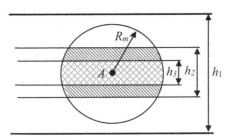

FIGURE 2.4 Origin of the disjoining pressure as understood from simple qualitative considerations. h_i, $i = 1.3$ depicts the possible thickness of the film.

molecule A will interact with the smaller number of molecules than a bulk molecule. This means that its potential energy will be larger than that of a molecule placed within the film placed in the film with the thickness h_1 (recall that the potential energy is negative, as shown in Figure 1.5). When the thickness of the liquid fill is h_3 (see Figure 2.4; the potential energy of molecules will be even larger and it will be much larger than the potential energy of the molecule located at the surface of the "thick" layer of the same liquid. The energy of molecules filling a thin film will be decreased when the thickness of the entire film is increased. The tendency to increase the thickness of the film results in the origin of the "strut off" pressure, which is called the disjoining pressure.[22]

Considering the disjoining pressure becomes important for very thin angstrom-scaled films; however, when the liquid is water, the range of the effects promoted by the disjoining pressure could be as large as 100 angstroms, due to the Helmholtz-charged double layer.[2,13] Actually, the disjoining pressure is built from three main components: the van der Waals interaction inspired term, the contribution due the electrical double layer and the so-called *structural component* caused by orientation of water molecules in the vicinity of the solid surface or at the aqueous solution/vapor interface.[13,19]

Disjoining pressure may be important in constituting the contact angle. Starov and Velarde suggested that the solid substrate is covered by a thin layer of a thickness e of absorbed liquid molecules[19] (see Figure 2.5). Considering the disjoining pressure introduced in the previous section gave rise to the following equation defining the contact angle θ:

$$cos\theta \cong 1 + \frac{1}{\gamma}\int_{e}^{\infty}\Pi(e)de, \qquad (2.14)$$

where $\Pi(e)$ is the aforementioned disjoining pressure (see Eq. (2.13)). Emergence of $\Pi(e)$ in Eq. (2.14) predicting the contact angle is natural; the thickness of the adsorbed liquid layer is supposed to be nano-scaled.[19] It should be stressed that the contact angle θ needs redefinition because the droplet cap does not touch the solid substrate, as shown in Figure 2.5. Starov and Velarde define the contact angle in this case as an angle between the horizontal axis and the tangent to the droplet cap profile at the point where it touches the absorbed layer of molecules (which is also called the *precursor film*).[16] Compare Eq. (2.14) with the Young equation (Eq. (2.9)); it is recognized that Eq. (2.14) does not contain the solid/liquid and solid air surface

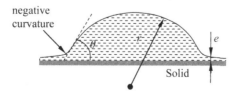

FIGURE 2.5 Droplet of the radius r surrounded by the thin layer of liquid of the thickness e governed by the disjoining pressure.

tensions. Eq. (2.14) tells us that the contact angle is defined by the liquid/air surface tension γ and the curve $\Pi(e)$ also called the "Derjaguin isotherm".[15–19] Thus, Eq. (2.14) presents very different approach to the prediction of the equilibrium contact angles than that emerging for the Young equation.

Let us estimate the disjoining pressure in the absorbed layer according to $|\Pi(e)| = \frac{A}{6\pi e^3}$, $A > 0$ (see Eq. (2.13)). If we will assume $A \approx 10^{-20} \div 10^{-19}$ J and $e = 1$ nm, we obtain giant values for the disjoining pressure: $|\Pi(e)| \cong 5 \times 10^4 \div 5 \times 10^5$ Pa. For $e = 10$ nm, we obtain much more reasonable values of the disjoining pressure: $|\Pi(e)| \cong 5.0 \div 50 \times 10^2$ Pa; however, they are still larger or comparable to the Laplace pressure in the drop. For $r \approx 1$ mm, we have $p = \frac{2\gamma}{r} \cong 140$ Pa (see Eq. (1.15)). How is the mechanical equilibrium possible in this case? Perhaps, it is due to the negative curvature of the droplet at the area where the cap touches the absorbed layer, shown in Figure 2.5. Moreover, if we take for the disjoining pressure $\Pi(e) > 0$ (which corresponds to $A < 0$, see Eq. (2.13)), we obtain from Eq. (2.14) $cos\theta \cong 1 + \frac{1}{\gamma}\int_e^\infty \Pi(e)de = 1 + \frac{A}{12\pi\gamma e^2} > 1$, which corresponds to complete wetting.[19] The latter condition implies that at oversaturation, no solution exists for an equilibrium liquid film thickness e outside the drop.[19]

In order to understand how the partial wetting is possible in this case, Starov and Velarde discussed more complicated forms of disjoining pressure, comprising the London-van der Waals and aforementioned double layer and structural contributions. They considered more complicated disjoining pressure isotherms, such as those depicted by curve 2 in Figure 2.6. The development of Eq. (2.14) yielded:

$$cos\theta \cong 1 + \frac{1}{\gamma}\int_e^\infty \Pi(e)de = 1 - \frac{S_- + S_+}{\gamma}, \qquad (2.15)$$

where S_- and S_+ are the areas depicted in Figure 2.6. Obviously (see Reference 16), the partial wetting is possible when $S_- > S_+$. Thus, when a droplet is surrounded by

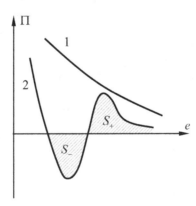

FIGURE 2.6 Disjoining pressure (Derjaguin's isotherms): 1 – isotherm corresponding to the complete wetting; only the London-van der Waals component is considered; 2 – isotherm comprising London, double layer and structural contributions and corresponding to the partial wetting.

a thin layer of liquid, the possibility of partial wetting depends according to Starov and Velarde on the particular form of the Derjaguin isotherm.[19]

2.5 CAPILLARY RISE

One of the most important and widespread wetting phenomena, resulting from the interplay of surface tension induced effects and gravity, is the rise (or descent) of liquid in capillary tubes, illustrated by Figure 2.7A–C. When a narrow tube is brought in contact with a liquid, some liquids (such as water in a glass tube) will rise and some (mercury in a glass tube) will descend into the tube.

Capillary rise is of a common occurrence in nature and technology. What is the physical reason for capillary rise? Let us consider an ideal (smooth, non-deformable, non-reactive) capillary tube wetted by a liquid. In tubes with an inner radius smaller than the capillary length l_{ca}, introduced in Section 1.11, the meniscus within a tube is

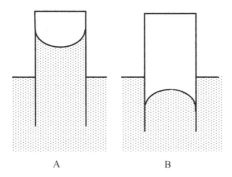

A   B

C

FIGURE 2.7 (A) Capillary rise: water in the glass tube; (B) capillary descent: mercury in the capillary tube; (C) water rise in glass capillary tubes.

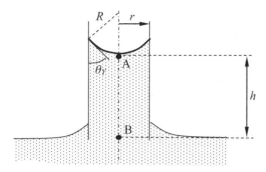

FIGURE 2.8 Capillary rise in a cylindrical tube: the Young contact angle is θ_Y.

a portion of a sphere. The radius of this sphere equals $R = \frac{r}{\cos\theta_Y}$, where r is the radius of the capillary tube (see Figure 2.8) and θ_Y is the contact angle of the *ideal tube/liquid* pair (Young equilibrium contact angle). The pressure in Point A (immediately underneath the meniscus) is given by $p_A = p_{atm} - \frac{2\gamma\cos\theta_Y}{r}$, where p_{atm} is atmospheric pressure. The pressure p_B in point B ($z = 0$) equals p_{atm}. However, $p_B - p_A = \rho g h$ (see Figure 2.8). Substituting p_B and p_A yields the well-known Jurin law:

$$h = \frac{2\gamma\cos\theta_Y}{\rho g r}. \tag{2.16}$$

It will be useful to rewrite Eq. (2.16) in the following form:

$$h = \frac{2 l_{ca}^2}{r}\cos\theta_Y. \tag{2.17}$$

strengthening the importance of the capillary length in the problems where the physics is defined by the interplay of surface tension and gravity.

Capillary rise is responsible for plenty of natural and technological phenomena; however, it is usually illustrated by an effect to which it is not related. It is a widespread myth that capillarity is responsible for the sap rise in tree capillaries. The ascent of sap in trees has long intrigued natural philosophers, including Leonardo da Vinci. Leonardo da Vinci suggested that in trees, the total cross-section area of branches is conserved across branching nodes.[23,24] This structure is believed to be self-similar and optimized to resist wind-induced loads.[23,24] What is the mechanism of the capillary rise: Francis Darwin stated that: "To believe that columns of water should hang in the tracheas like solid bodies, and should, like them, transmit downwards the pull exerted on them at their upper ends by the transpiring leaves, is to some of us equivalent to believing in ropes of sand".[25]

What is the mechanism of the sap rise in the trees (which may be as high as 100 m typical for *Sequoia*, popularly known as redwood trees)? The orthodox cohesion-tension theory, relating to the sap rise in the trees, is formulated as follows: first, that trees sap forms a myriad of broken and more importantly unbroken threads, stretching from the absorbing surfaces of the roots to the evaporation surfaces of the leaves, through the vascular structure of the xylem in the stem. Second, that transpiration principally from the leaves is the trigger of the driving force behind the ascent of sap. Transpiration, or the removal of water molecules from the water–air interfaces (menisci) within the pores, causes the interfaces to recede into the pores and change their shape (become more concave). The combination of forces of adhesion to the cell walls and surface tension on the interfaces (i.e., capillarity) acts to restore the equilibrium shape of the menisci and thereby creates tension (negative pressure) in the sap in their vicinity.

Let us estimate the maximal capillary rise according to Eqs. (2.16) and (2.17) if the complete wetting of capillary vessels is assumed, i.e., $\cos\theta_Y = 1$. The characteristic radius of capillary vessels in trees is close to 10 μm (Reference 26). Substituting $\gamma_w = 70.0\,\frac{mJ}{m^2}$; $\rho_w \cong 10^3\,\frac{kg}{m^3}$; $r \cong 10^{-5}$ m, into Eq. (2.16), we obtain for the most optimistic estimation of the maximal water rise in tree capillary vessels $h \cong 1.4$ m. At the same time, water is transported even to 100-m height redwood trees. Thus, it was concluded that "standard arguments in favor of the so-called Cohesion-Tension (i.e. capillary) theory, the conventional wisdom concerning the remarkable mechanism behind the ascent of sap in trees, are completely misleading" (for the extended discussion of the problem, see References 25–27). The mechanism of water rise in trees is not understood today to a full extent; however, it is generally accepted that water is pulled from the roots to the leaves by a pressure gradient arising from evaporation of water from the leaves. Negative pressures as high as −100 atm were registered in plants.[27,28]

Capillary rise can be also observed when liquid is confined between two vertical planes separated by a distance w, as shown in Figure 2.9. In the case of *ideal planes*, the Laplace pressure is given by $p_L = \gamma\left(\frac{1}{R_1} + \frac{1}{R_2}\right) = \frac{\gamma}{R} = \frac{2\gamma\cos\theta_Y}{w}$ (assume that the shape of the meniscus is a cylinder and consider $R_2 = \infty$; $R_1 = R = \frac{w}{2\cos\theta_Y}$). Calculations akin to those leading to Eqs. (2.16) and (2.17) yield:

$$h = \frac{2\gamma\cos\theta_Y}{\rho g w} = 2\frac{l_{ca}^2}{w}\cos\theta_Y.\qquad(2.18)$$

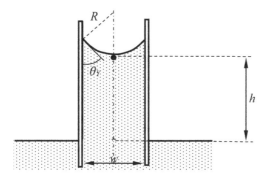

FIGURE 2.9 Capillary rise between two vertical ideal plates. The separation between plates is w.

2.6   WETTING OF REAL SURFACES—CONTACT ANGLE HYSTERESIS: HIS MANIFESTATIONS AND ORIGIN

Now we come to the very important theme, crucial for understanding wetting phenomena and in particular the Lotus effect. The Young equation given by Eq. (2.3), i.e., $cos\theta_Y = \frac{\gamma_{SA}-\gamma_{SL}}{\gamma}$ predicts a sole value of the contact angle for a given ideal solid/liquid pair. As it always occurs in reality, however, the situation is much more complicated and rich in its physical content. Let us deposit a droplet onto an inclined plane, as described in Figure 2.10 in the situation of partial wetting (the spreading parameter $\Psi < 0$).

The inclined plane is supposed to be ideal, i.e., atomically flat, chemically homogeneous, isotropic, insoluble, non-reactive and non-deformed. We will nevertheless recognize different contact angles θ_1, θ_2 as shown in Figure 2.10. This experimental observation definitely contradicts the predictions of the Young equation. Moreover, a droplet on an inclined plane could be in equilibrium only when contact angles θ_1, θ_2 are different.[2,3] If we increase the inclination angle α, contact angles θ_1, θ_2 will change, and at some critical angle α, the droplet will start to slip. This critical contact angle is called the *sliding angle*. We conclude that a variety of contact angles can be observed for the same ideal solid substrate/liquid pair. It turns out that an accurate physical treatment of the droplet behavior on the inclined surface presents extremely challenging physical problem, which is not completely solved until now. Excellent modern review of the state-of-the-art in the field is supplied in Reference 29.

Let us perform one more simple experiment. When a droplet is inflated with a syringe as shown in Figure 2.11A, we observe the following picture: the triple line is pinned to the substrate up to a certain volume of the droplet. When the triple line is pinned, the contact angle increases till a certain threshold value θ_A beyond which the triple line does move. The contact angle θ_A is called the *advancing contact angle*.[2,3,29] When a droplet is deflated as depicted in Figure 2.11B, its volume can be decreased to a certain limiting value; in parallel, the contact angle decreases till a threshold value θ_R, known as the *receding contact angle*.[2,3,29] When $\theta = \theta_R$, the triple line suddenly moves. Both θ_A and θ_R are equilibrium, although metastable contact angles.[2,3,29] The difference between θ_A and θ_R is called the *contact angle hysteresis*.[2,3,29–31]

One more manifestation of the contact angle hysteresis is presented in Figure 2.12. Actually, this effect is well-known to most people: a vertical column of liquid placed

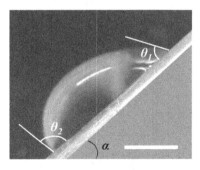

FIGURE 2.10   Drop on the inclined plane. Difference between contact angles θ_1, θ_2 prevents the droplet sliding. α is the inclination angle. Scale bar is 3 mm.

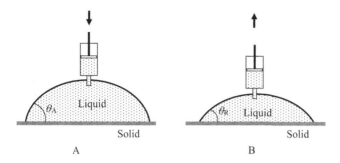

FIGURE 2.11 Inflating and deflating of a droplet. Advancing θ_A (**A**) and receding contact angles θ_R (**B**) are shown.

into a vertical tube does not fall but is retained by molecular interaction between molecules of the tube and those of the liquid, giving rise to deformation of the liquid surface and resulting in capillary menisci. The difference between the contact angle at the lower and upper menisci makes possible the balance of forces:

$$\frac{2\gamma}{r}(\cos\theta_1 - \cos\theta_2) = \rho gH. \tag{2.19}$$

The maximal high of the liquid column H_{max}, which could be retained by the capillary tube, is given by

$$\frac{2\gamma}{r}(\cos\theta_R - \cos\theta_A) = \rho gH_{max}, \tag{2.20}$$

where θ_A and θ_R are the receding and advancing contact angles, respectively.

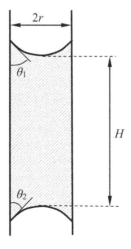

FIGURE 2.12 Manifestation of the contact angle hysteresis in the capillary tube: the column of liquid is retained by the contact angle hysteresis.

Let us notice one regrettably widespread misunderstanding in the literature devoted to the wetting phenomena. The advancing and receding contact angles are mentioned as "dynamic contact angles", and this is exactly a mistake. The advancing and receding contact angles are essentially static equilibrium contact angles.

Both measurement and understanding of the phenomenon of the contact angle hysteresis remain challenging experimental and theoretical tasks. Experimental establishment of the receding contact angle, regrettably, depends strongly on the experimental technique involved in the measurement and the reported values of the receding contact angles are poorly reproducible (advancing contact angles, contrastingly, are highly reproducible).[32] It is customary to attribute the phenomenon of the contact angle hysteresis to physical or chemical heterogeneities of the substrate[2,3]; however, even ideal substrates discussed in Section 2.2 demonstrate significant contact angle hysteresis. We'll start our discussion from the physical reasons of the contact angle hysteresis on ideal substrates.

2.7 CONTACT ANGLE HYSTERESIS ON SMOOTH HOMOGENEOUS SUBSTRATES

Contact angle hysteresis has been registered even for silicon wafers which are regarded as atomically flat rigid substrates and are considered very close to be ideal ones. C. Extrand studied the contact angle hysteresis of various liquids, including water, ethylene glycol, methylene iodide, acetophenone and formamide, deposited on silicon wafers with a tilted plane method.[33] Contact angle hysteresis (defined as $\Delta\theta = \theta_A - \theta_R$) as high as $14°$ was established for the water/silicon wafer and methylene iodide/silicon wafer pairs. It should be mentioned that the contact angle hysteresis on the order of magnitude of $5–10°$ has been reported for other silicon wafer/liquid pairs.[33] High contact angle hysteresis has been observed also for atomically smooth polymer substrates. Lam et al. used polymer-coated silicon wafers for the study of the contact angle hysteresis and reported the values of contact angle hysteresis on the order of tens of degrees.[34] The question is: how is such dispersion of contact angles possible, in contradiction to the predictions of the Young equation?

The explanation for the contact angle hysteresis observed on smooth surfaces becomes possible if we consider the effect of the *pinning of the triple line*. The intermolecular forces acting between molecules of solid and those of liquid, which pin the triple line to the substrate, are responsible for the contact angle hysteresis. Yaminsky developed an extremely useful analogy between the phenomena occurring at the triple line with the static friction.[35] I quote: "...for a droplet on a solid surface there is a static resistance to shear. It occurs *not over the entire solid-liquid interface, but only at the three-phase line* ...This paradox is easily resolved once one realizes that the liquid-solid interaction is in fact not involved in the process of overflow of liquids above solid surfaces. A boundary condition of zero shear velocity typically occurs even for liquid-liquid contacts ... But even given that the strong binding condition does apply to solid-liquid interfaces, this does not prevent the upper layer of the liquid from flowing above the 'stagnant layer' of a gradient velocity. The movement of the liquid over the wetted areas occurs in the absence of static resistance. Interaction in a manner of dry friction occurs only at the three-phase line".[35]

Thus, the contact angle hysteresis on ideal surfaces is caused by the intermolecular interaction between molecules constituting a solid substrate and a liquid; this interaction pins the triple line and gives rise to a diversity of experimentally observed contact angles. The novel concept explaining and predicting the phenomenon of the contact angle hysteresis is the concept of adaptive wetting, introduced in Reference 36.

Many surfaces reversibly change their structure and interfacial energy upon being in contact with a liquid. Such surfaces adapt to a specific liquid. The kinetic model to describe contact angles of such adaptive surfaces was suggested in Reference 36. By introducing exponentially relaxing interfacial energies (implying that the solid surface energy decreases exponentially with time as a result of contact with a new fluid) and applying Young's equation locally, the authors predicted a change of advancing and receding contact angles depending on the velocity of the contact line. It should be emphasized that even for perfectly homogeneous and smooth surfaces, a dynamic contact angle hysteresis was predicted. From the microscopic point of view, the adaptive wetting is seen as re-orientation of functional groups present at the solid/liquid interface.[29,36]

2.8   CONTACT ANGLE HYSTERESIS ON REAL SURFACES

Real surfaces are rough and chemically heterogeneous. The macroscopic parameter describing wetting of surfaces is the *apparent contact angle*. The apparent contact angle is an equilibrium contact angle measured macroscopically on a solid surface that may be rough or chemically heterogeneous, or in other words is the contact angle measured experimentally on the macroscopic scale, as it is stated in Reference 37, devoted to the development of accurate terminology devoted to the wetting phenomena. The apparent contact angle is the only one that can be routinely measured.[37]

The detailed microscopic topography of a rough or chemically heterogeneous surface cannot be viewed with regular optical means; therefore, this contact angle is defined as the angle between the tangent to the liquid-vapor interface and the apparent solid surface as macroscopically observed.[2,3,37] Actually a spectrum of apparent contact angles is observed on real surfaces. A diversity of physical factors contributes to the contact angle hysteresis, including the pinning of the triple line, liquid penetration and surface swelling, deformation of the substrate, etc.[29–31,37,38] It should be emphasized that the contact angle hysteresis turned out to be a complicated, time-dependent effect. As seen from the phenomenological point of view, the contact angle hysteresis is due to the multiple minima of the free energy of a droplet deposited on the substrate.[39] These minima are separated by potential barriers.[39] Contact angle hysteresis depends on the chemical heterogeneity and roughness on the surface which, in turn, influence the pinning of the triple line.[31] Contact angle hysteresis may be strengthened and weakened by the roughness of a substrate. It is high for the Wenzel-like wetting and low for the Cassie-like wetting to be discussed further in detail. It is often stated erroneously in the literature that the *superhydrophobic materials* are those which demonstrate the apparent contact angles close to 180°, as shown in Figure 2.1C. Actually, a high apparent contact angle is a *necessary but insufficient condition of superhydrophobicity*. Low contact angle hysteresis providing easy transportation of droplets along the surface and the high stability of the Cassie wetting regime are crucially necessary for a true superhydrophobicity and self-cleaning. A comprehensive review of the contact angle hysteresis is supplied in Reference 31.

2.9   THE DYNAMIC CONTACT ANGLE

Until now, we have discussed only the statics of wetting. Now we'll consider a much more complicated situation: when the triple line moves.[37] And this is the case when a droplet bounces the solid substrate, which is obviously important for industrial applications of superhydrophobic surfaces. The triple line in this situation is displaced. Generally speaking, this movement may arise from the motion of the substrate or the droplet itself. When the triple line moves, the dynamic contact angle θ_D, does not equal the Young angle as shown in Figures 2.13 and 2.14.

It can be larger or smaller than the Young angle (see Figures 2.13 and 2.14). The excess force pulling the triple line is given by (Reference 2):

$$F(\theta_D) = \gamma_{SA} - \gamma_{SL} - \gamma \cos\theta_D. \tag{2.21}$$

As we already mentioned in the previous section, the effect of contact angle hysteresis complicates the study of wetting even in the simplest possible static situation. The movement of the triple line introduces additional difficulties, so reproducing the results of the measurements of dynamic contact angles becomes a challenging task. We'll start from the theoretical analysis of dynamic wetting on ideally smooth, rigid, non-reactive surfaces.

Now we find ourselves in the realm of hydrodynamics. Systematic study of the problem of the dynamics of wetting has been undertaken by Voinov.[40] Thus, the

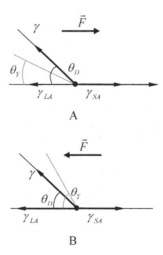

FIGURE 2.13   Origin of the dynamic contact angle: (**A**) The dynamic contact angle θ_D is larger than the Young angle θ_Y. (**B**) The opposite situation: the dynamic contact angle θ_D is smaller than the Young angle θ_Y.

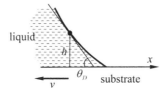

FIGURE 2.14 Formation of the dynamic contact angle θ_D.

dissipative processes, depending on the viscosity of the liquid denoted η, introduced in Section 1.4 become important. Voinov addressed the situation, when the inertia-related contributions are neglected; this corresponds to the cases, when $Re \ll 1$ takes place, where Re is the Reynolds number $Re = \frac{\rho v D}{\eta}$, introduced into Section 1.15 by Eq. (1.22) (D is the characteristic size of a droplet, which may be equal to its diameter, and v is the velocity of the substrate). Thus, only dimensionless number, governing the flow, is the *capillary number Ca*, defined by Eq. (1.19) as $Ca = \frac{\eta v}{\gamma}$. Voinov also phenomenologically introduced the angle of the free surface slope θ_m at the height of the limiting scale h_m (see Figure 2.14):

$$\theta_D = \theta_m; h = h_m. \tag{2.22}$$

Voinov noted that θ_m is unknown beforehand and should be determined during the solution of the problem.[40] The accurate mathematical solution of the hydrodynamic problem of a dynamic wetting yielded for the dynamic contact angle:

$$\theta_D(h) = \left[\theta_m^3 + 9\frac{\eta v}{\gamma}\ln\frac{h}{h_m}\right]^{1/3} = \left[\theta_m^3 + 9Ca\ln\frac{h}{h_m}\right]^{1/3}. \tag{2.23}$$

Formula (2.23) is referred as the Cox-Voinov law, and it is valid for $\theta_D < \frac{3\pi}{4}$ (References 27 and 28). Hoffmann has shown that the experimental dependence $\theta_D(Ca)$ is represented by a universal curve (corrected with a shifting factor) for a diversity of liquids.[41] It is seen from Eq. (2.23) that the slope varies logarithmically with the distance from the triple line. Thus, it is impossible to assign a unique dynamic contact angle to a triple line moving with a given speed.[42] Hence, Figure 2.13 depicts an obvious oversimplification of the actual dynamic wetting situation. It is also noteworthy that θ_D depends slightly on the cutoff length h_m; however, it depends strongly on the microscopic angle θ_m. And again, the dynamic contact angle, predicted by Eq. (2.23), is independent of gravity.[43]

2.10 WETTING OF HETEROGENEOUS SURFACES: THE WENZEL MODEL

Wetting of so-called real surfaces, which may be rough or heterogeneous, is described by two main physical models, namely the Wenzel and Cassie wetting models. In this section, we develop the basic model describing the wetting of rough, however chemically homogeneous surfaces, i.e., the Wenzel wetting model. The Wenzel models

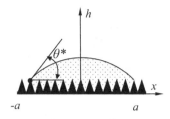

FIGURE 2.15 The Wenzel wetting of a chemically homogeneous rough surface: liquid completely wets the grooves.

predict the apparent contact angle, which is an essentially macroscopic parameter. This fact limits the field of validity of these models: they work when the characteristic size of a droplet is much larger than that of the surface heterogeneity or roughness. The use of the Wenzel and Cassie equations needs a certain measure of care; numerous misinterpretations of these models are found in the literature. We'll discuss the applicability of these basic models in detail.

The Wenzel model, introduced in 1936, deals with the wetting of rough, chemically homogeneous surfaces and implies total penetration of a liquid into the surface grooves, as shown in Figure 2.15.

When the spreading parameter $\Psi < 0$ (see Section 2.1), a droplet forms a cap resting on the substrate with an apparent contact angle θ^*. The interrelation between the apparent and the Young contact angles is given by the Wenzel equation:

$$cos\theta^* = \tilde{r} cos\theta_Y, \tag{2.24}$$

where $\tilde{r}$ is the roughness ratio of the wet area: in other words, the ratio of the real surface in contact with liquid to its projection onto the horizontal plane; obviously, $\tilde{r}$ is dimensionless. Parameter $\tilde{r} > 1$ describes the increase of the wetting surface due to roughness. Formula (2.24) presents the famous *Wenzel equation*.[1] Three important conclusions follow from Eq. (2.24):

- Inherently smooth hydrophilic surfaces ($\theta_Y < \frac{\pi}{2}$) will be more hydrophilic when riffled: $\theta^* < \theta_Y$ due to the fact that $\tilde{r} > 1$.
- Due to the same reason, inherently hydrophobic flat surfaces ($\theta_Y > \frac{\pi}{2}$) will be more hydrophobic when grooved: $\theta^* > \theta_Y$.

The Wenzel angle given by Eq. (2.24) is *independent of the droplet shape and external fields U* under very general assumptions discussed in detail in References 3 and 4. Surfaces characterized by the Wenzel-like, homogeneous wetting are usually characterized by a high contact angle hysteresis; this fact is extremely important for the development of the self-cleaning, superhydrophobic surfaces, demanding the lowest possible contact angle hysteresis. Consider that for the practical use of the Wenzel Eq. (2.24), the Young contact angle, established on the ideal (flat, non-reactive, non-deformable) surface, should be independently measured. For the accurate derivation and extensions of Eq. (2.24), see References 3 and 4.

2.11   THE CASSIE-BAXTER WETTING MODEL

The Cassie-Baxter wetting model introduced in References 44 and 45 deals with
the wetting of *flat chemically heterogeneous* surfaces. Suppose that the surface
under the drop is flat but consists of *n* sorts of materials randomly distributed
over the substrate as shown in Figure 2.16. This corresponds to the assumptions
of the Cassie-Baxter wetting model.[44,45] Each material is characterized by its
own surface tension coefficients $\gamma_{i,SL}$ and $\gamma_{i,SA}$ (SL and SA denote solid-liquid
and solid-air interfaces), and by the fraction f_i in the substrate surface, the condition $\Sigma_{i=1}^{n} f_i = 1$ takes place. The apparent contact angle θ^* in this situation is
given by

$$cos\theta^* = \frac{\sum_{i=1}^{n} f_i \left(\gamma_{i,SA} - \gamma_{i,SL} \right)}{\gamma}, \tag{2.25}$$

predicting the so-called Cassie apparent contact angle θ^* on flat chemically heterogeneous surfaces (for the rigorous variational analysis based grounding of Eq.
(2.25), see References 3 and 4). It was demonstrated that the Cassie apparent contact angles are also insensitive to external fields, including gravity.[3,4] When the
substrate consists of two kinds of species, the Cassie-Baxter equation obtains the
simple form:

$$cos\theta^* = f_1 cos\theta_1 + f_2 cos\theta_2 \tag{2.26}$$

which is widespread in the scientific literature dealing with the wetting of heterogeneous surfaces.[45] For an accurate interpretation extensions of the Cassie-Baxter
model to curved surfaces and situations when the line tension is essential (see
Section 2.3), see References 3 and 46–48. The Cassie wetting regime enables manufacturing of the surfaces with a low contact angle hysteresis (see Sections 2.6–2.8),
which is extremely important for development of the self-cleaning materials. It
should be stressed that the Wenzel and Cassie-Baxter apparent contact angles are
equilibrium ones. Their experimental establishment remains problematic due to the
effect of the contact angle hysteresis.

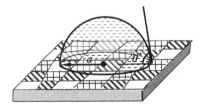

FIGURE 2.16   The Cassie-Baxter wetting of flat chemically heterogeneous surfaces (various colors correspond to different chemical species). *a* is the radius of the droplet; θ^* is the
Cassie-Baxter apparent contact angle.

2.12 CASSIE-BAXTER WETTING IN A SITUATION WHERE A DROPLET PARTIALLY SITS ON AIR

The peculiar form of the Cassie-Baxter equation given by Eq. (2.26) was successfully used for the explaining the phenomenon of superhydrophobicity, which will be discussed in detail in Section 2.16. Jumping ahead, we admit that in the superhydrophobic situation, a droplet is partially supported by solid substrate and partially by air cushions, as shown in Figure 2.17.

Consider a situation where the mixed surface comprises solid surface and air pockets, with the contact angles θ_Y (which is the Young angle of the solid substrate) and π, respectively. We denote by f_S and $1 - f_S$ relative fractions of solid and air, respectively. Thus, we deduce (from Eq. (2.26)):

$$cos\theta^* = -1 + f_S\left(cos\theta_Y + 1\right). \tag{2.27}$$

Eq. (2.27) predicts the apparent contact angle θ^* in the situation where a droplet sits partially on solid and partially on air, and it was shown experimentally that it does work for a diversity of porous substrates.[2,3,47] It is noteworthy that switching from Eq. (2.26) to Eq. (2.27) is not trivial and straightforward, because the triple (three-phase) line could not be at rest on pores.[11,49] For the extended discussion of applicability of Eq. (2.27), see Reference 49. It is seen from Eq. (2.27) that when $f_S \to 0$, we predict $cos\theta^* \to -1$ and, respectively, the spreading parameter, introduced into Section 2.1, $\Psi \to -2\gamma$. Thus, we obtain the situation of superhydrophobicity or "lotus effect", illustrated in Figures 2.1C and 2.18.

2.13 CASSIE-BAXTER IMPREGNATING WETTING

There exists one more possibility of the heterogeneous wetting: this is the so-called Cassie-Baxter impregnating wetting state first introduced in Reference 50 and reported experimentally in References 51–53. In this case, liquid penetrates into the grooves of the solid and the drop finds itself on a substrate viewed as a patchwork of solid and liquid (solid "islands" ahead of the drop are dry, as shown in Figure 2.19). Thus, we deal with a heterogeneous wetting in this case, namely a droplet is supported by the heterogeneous as shown in Figure 2.19.

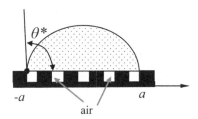

FIGURE 2.17 The particular case of the Cassie wetting: a droplet is partially supported by solid and partially by air cushions.

FIGURE 2.18 The phenomenon of superhydrophobicity is shown. A 50-μl water droplet deposited on a pigeon feather. The pronounced superhydrophobicity of the feather is clearly seen; the apparent contact angle of a water droplet is close to π; water is easily sliding of the feather. The Cassie wetting state is thermodynamically stable. Scale bar is 3 mm.

This wetting state should be distinguished from the Wenzel wetting illustrated in Figure 2.15. When the Wenzel wetting occurs, the *solid outside of the triple line is dry*, whereas in the Cassie-Baxter impregnating situation, it is partially wetted by liquid, as shown in Figure 2.19. The Cassie-Baxter Eq. (2.26) can be applied to the mixed surface depicted in Figure 2.19, with contact angles θ_Y and zero, respectively. We then derive for the apparent contact angle θ^*:

$$cos\theta^* = 1 - f_S + f_S cos\theta_Y. \qquad (2.28)$$

We denote by f_S and $1 - f_S$ the relative fractions of the solid and liquid phases underneath the droplet. Eq. (2.28) may be obtained from the variational principles

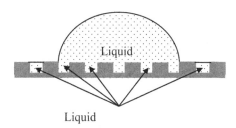

Liquid

FIGURE 2.19 The Cassie-Baxter impregnating wetting state is shown. Droplet finds itself on a substrate viewed as a patchwork of solid and liquid; solid "islands" ahead of the drop are dry.

for the composite surface which comprises two species, characterized by the Young angles of θ_Y and zero. As it was demonstrated in References 2 and 14, the Cassie-Baxter impregnating wetting is possible when the Young angle satisfies Eq. (2.29):

$$cos\theta_Y > \frac{1-f_s}{\tilde{r}-f_s},\tag{2.29}$$

where $\tilde{r}$ and f_s are the aforedefined parameters of the Wenzel and Casse-Baxter wetting models. As shown in Reference 51, the Cassie-Baxter impregnating state corresponds to the lowest apparent contact angle θ^* for a certain solid/liquid pair, when compared to that predicted by the Wenzel (Eq. (2.24)) and the Cassie-Baxter air trapping (Eq. (2.27)) wetting regimes. The Cassie-Baxter impregnating state becomes important in a view of wetting transitions on rough surfaces discussed further in Section 2.19. It should be emphasized that the possibility of the Cassie-Baxter impregnating state is usually underestimated in the scientific literature; it is widespread delusion that only Cassie air trapping and Wenzel wetting state exist.

2.14   THE IMPORTANCE OF THE AREA ADJACENT TO THE TRIPLE LINE IN THE WETTING OF ROUGH AND CHEMICALLY HETEROGENEOUS SURFACES

At the first glance, it seems that the Cassie and Wenzel wetting models, discussed in the previous section, are rather intellectually transparent. However, in 2007, Gao and McCarthy initiated a stormy scientific discussion with their provocatively named paper, "How Wenzel and Cassie were wrong?", followed in 2009 by the paper "An attempt to correct the faulty intuition perpetuated by the Wenzel and Cassie Laws".[54,55] They put forward the following question: What will be the apparent contact angle in the situation presented in Figure 2.20, when a drop of a radius a is deposited on a flat surface comprising a spot of radius b which is smaller than the radius of the droplet? The substrate and the spot are made from different materials

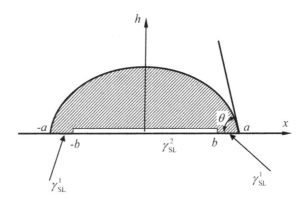

FIGURE 2.20   A drop of a radius a deposited axisymmetrically on a composite surface, comprising a "spot" with a radius b.

possessing various surface energies. The question is: Will this spot affect the contact angle? On the one hand, the surface is chemically heterogeneous and it seems that the spot will influence the contact angle; on the other hand, the intuition relating the Young equation to the equilibrium of forces acting on the triple line suggests that the contact angle will "feel" only the areas adjacent to the triple line, and the central spot will have no impact on the contact angle. The question may be generalized: Is the wetting of a composite surface a 1D or 2D affair? Or in other words: Is the apparent contact angle governed by the entire surface underneath a drop (2D scenario), or it is dictated by the area adjacent to the triple (three-phase) line (1D scenario)? The problem was cleared up in a series of papers.[56–59] The most general answer may be obtained within the variational approach developed in Reference 60.

Consider a liquid drop of a radius a deposited on a two-component composite flat surface including a round spot of a radius b (i.e., chemical heterogeneity) in the axisymmetric way depicted in Figure 2.20. The variational analysis carried out in Reference 60 demonstrated that the spot far from the triple line has no influence on the contact angle, and therefore a discrepancy with the force-based approach is avoided.

Now the most delicate point has to be considered. The question is: What is the precise meaning of the expression "far from the spot"? From the physical point of view, it means that the macroscopic approach is valid when a three-phase line is displaced, namely, $a - b \geq 100$ nm (see Figure 2.20); when this condition is fulfilled, particles located on the triple line do not "feel" the spot, i.e., the influence of the van der Waals forces is negligible (see Sections 1.5 and 2.4). It should be stressed that the apparent contact angle is essentially a macroscopic notion; hence, all our discussion assumes the macroscopic approach. At the same time, the area far from the triple line may contribute essentially to the adhesion of a droplet to the substrate.[61] An accurate mathematical treatment combining the surface energy minimization with the concept of deficit curvature from the Gauss-Bonnet was reported in Reference 62.

2.15   THE MIXED WETTING STATE

As it always takes place in nature, the pure Wenzel and Cassie wetting regimes introduced in previous sections are rare in occurrence. More abundant is a so-called mixed wetting state, depicted schematically in Figure 2.21, introduced in Reference 63 and discussed in much detail in Reference 64. In this situation, the droplet is supported partially by air and partially by *a rough chemically homogeneous solid surface*. The apparent contact angle in this case is given by

$$cos\theta^* = \tilde{r}f_S cos\theta_Y + f_S - 1. \qquad (2.30)$$

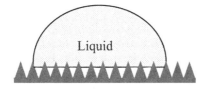

FIGURE 2.21   The mixed wetting state.

Obviously for $\tilde{r} = 1$, we return to the usual Cassie air trapping Eq. (2.27). Eq. (2.30) was derived first in Reference 63 and analyzed in Reference 64 and is extremely useful for understanding the phenomenon of superhydrophobicity to be discussed in detail in Section 2.16. The rigorous thermodynamic grounding of Eq. 2.30 may be found in Reference 3.

More accurate approach considering the effects due to the line tension (see Section 2.3) was developed in References 65 and 66. The role of the line tension in constituting apparent contact angles remains debatable, owing to the fact the value of to the line tension is not well established experimentally (see the discussion in Section 2.3 and References 5–7). It should be stressed that the apparent contact angles, predicted by the Wenzel, Cassie-Baxter and the "mixed wetting" models, are independent of external fields, volume and shape of droplets.[3]

2.16 SUPERHYDROPHOBICITY

The phenomenon of superhydrophobicity was revealed in 1997, when W. Barthlott and C. Neinhuis studied the wetting properties of a number of plants and stated that the "interdependence between surface roughness, reduced particle adhesion and water repellency is the keystone in the self-cleaning mechanism of many biological surfaces".[67] They discovered the extreme water repellency and unusual self-cleaning properties of the "sacred lotus" (*Nelumbo nucifera*) and coined the notion of the y and unusual self-cleaning properties of the "sacred lotus" (nhuis studied the wetting properties of a number of plants and stated that the "interdependence between surface p correlation between the surface roughness of plants, their surface composition and their wetting properties (varying from superhydrophobicity to superhydrophilicity).[31–33]

The amazing diversity of the surface reliefs of plants observed in nature was reviewed in References 68–71. Barthlott et al. noted that plants are coated by a protective outer membrane coverage, or *cuticle*. This cuticle is a composite material built up by a network of polymer *cutin* and waxes.[31–33] One of the most important properties of this cuticle is *hydrophobicity* which prevents the desiccation of the interior plants cells.[31–33] It is noteworthy that the cuticle demonstrates only moderate inherent hydrophobicity (or even *hydrophilicity* for certain plants such as the famous lotus (Reference 34)), whereas the rough surface of the plant may be extremely water repellent.

Barthlott et al. also clearly understood that the micro- and nano-structures of the plants surfaces define their eventual wetting properties, in accordance with the Cassie-Baxter and Wenzel models (discussed in detail in the previous sections). Since Barthlott et al. reported the extreme water repellency of the lotus, similar phenomena were reported for a diversity of biological objects: rice leaves,[72–74] butterfly wings[75] water strider legs[76,77] as well as bird wings, shown in Figure 2.18.[78] It is noteworthy that the keratin constituting bird wings is also inherently *hydrophilic*.[78] Barthlott et al. also drew the attention of investigators to the *hierarchical reliefs* inherent to plants characterized by superhydrophobicity, such as depicted in Figure 2.22. The interrelation between the hierarchical topography of surfaces and their water repellency will be discussed below in detail.

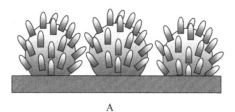

A

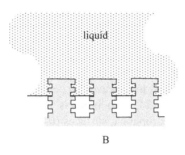

B

FIGURE 2.22 Typical hierarchical reliefs inherent to lotus-like surfaces. (**A**) Natural hierarchical surface. Two spatial scales are recognized. (**B**) Hydrophobic pillars possessing rough side facets, increasing energy barrier, separating the Cassie and Wenzel wetting states. Again, two spatial scales appear in the relief.

2.17 SUPERHYDROPHOBICITY AND THE CASSIE-BAXTER WETTING REGIME

In this section, we deal with the wetting of micro- or nano-rough surfaces. The wetting of these surfaces is characterized by an *apparent contact angle*, introduced in Section 2.8. The surfaces characterized by an apparent contact angle larger than 150° are referred to as superhydrophobic.[68–70] It should be immediately emphasized that high apparent contact angles observed on a surface are not sufficient for referring it as superhydrophobic. True superhydrophobicity should be distinguished from the pseudo-superhydrophobicity inherent to surfaces exhibiting the "rose petal effect" to be discussed later. The pseudo-superhydrophobic surfaces are characterized by large apparent contact angles accompanied by the high contact angle hysteresis, discussed in great detail in Sections 2.7 and 2.8. In contrast, truly superhydrophobic surfaces are characterized by large apparent contact angles and low contact angle hysteresis resulting in a low value of a sliding angle: a water drop rolls along such a surface even when it is tilted at a small angle. Truly superhydrophobic surfaces are also *self-cleaning*, since rolling water drops wash off contaminations and particles such as dust or dirt. Actually, the surface should satisfy one more demand to be referred to as superhydrophobic: the Cassie-Baxter wetting regime on this surface should be thermodynamically stable. The stability of the Cassie-Baxter wetting regime is important for preventing the Cassie-Wenzel wetting transitions (to be discussed in Section 2.19).

The Cassie-Baxter Eq. (2.27), developed for the air trapping situation where the droplet is partially supported by air cushions (see Figure 2.17), supplies the natural explanation for the phenomenon of superhydrophobicity. Indeed, the apparent contact angle θ^* in this situation given by $cos\theta^* = -1 + f_S(cos\theta_Y + 1)$ ultimately approaches π when the relative fraction of the solid f_S approaches zero. This corresponds to complete dewetting, discussed in Section 2.1 and illustrated in Figure 2.1C. Note that the apparent contact angle also approaches π when the Young angle tends to π. However, this situation is practically unachievable, because the most hydrophobic flat polymer, polytetrafluoroethylene (Teflon), demonstrates an advancing angle smaller than 120°, and this angle is always larger than the Young one. Hence, it is seen from the Cassie-Baxter equation that the apparent contact angles could be increased by decreasing the relative fraction of the solid surface underneath a droplet. However, there exists a more elegant way to manufacture surfaces characterized by ultimately high apparent contact angles: producing hierarchical reliefs, and this is the situation observed in natural objects such as lotus leaves (to be discussed in the next section).

Note that the Wenzel Eq. (2.24) also predicts high apparent contact angles approaching π for inherently hydrophobic surfaces ($\theta_Y > \frac{\pi}{2}$), when $\tilde{r} \gg 1$. However, the Wenzel-like wetting, depicted in Figure 2.15, is characterized by the high contact angle hysteresis, whereas superhydrophobicity accompanied by self-cleaning calls for the contact angle hysteresis to be as low as possible.

2.18 WETTING OF HIERARCHICAL RELIEFS

Maximal air trapping providing high apparent contact angles is provided by the so-called hierarchical reliefs, as depicted in Figure 2.22.

Herminghaus developed a very general approach to the wetting of hierarchical reliefs, based on the concept of the effective surface tension of a rough solid/liquid interface.[79] For hierarchically indented substrates, Herminghaus deduced the following recursion relation:

$$cos\theta_{n+1} = (1 - f_{Ln})cos\theta_n - f_{Ln}, \qquad (2.31)$$

where n denotes the number of the generation of the indentation hierarchy and where f_{Ln} is the fraction of free liquid surfaces suspended over the indentations of the relief of nth order, A larger n corresponds to a larger length scale. According to Eq. (2.31), $cos\theta_{n+1} - cos\theta_n = -f_{Ln}(1 + cos\theta_n) < 0$, so that the sequence represented by Eq. (2.31) is monotonic. Thus, addition of the scale to existing hierarchy will necessarily increase the apparent contact angle. Herminghaus stressed that θ_0 corresponding to the Young contact angle θ_Y must only be finite, but need not exceed $\frac{\pi}{2}$ for obtaining high resulting apparent contact angles on hierarchical surfaces.[79] This result explains the possibility of creation of superhydrophobic surfaces from inherently hydrophilic materials. This possibility is of a primary importance for materials science engineering, enabling design of superhydrophobic materials such as metals and ceramics.[80–82] Excellent review of the methods enabling manufacturing of superhydrophobic topographies from metals is supplied in Reference 80. Herminghaus also considered fractal surfaces and estimated the Hausdorf dimension of such surfaces.[79] Generally, the model proposed by Herminghaus successfully explained high apparent contact angles observed on a diversity of biological objects.[67–78]

Herminghaus discussed a very general situation of wetting of fractal hierarchical structures.[79] Actually, both natural and artificial superhydrophobic surfaces are usually built of twin-scale surfaces, such as discussed in References.[80–82]

2.19 WETTING TRANSITIONS ON ROUGH SURFACES

We already mentioned that high apparent contact angles and low contact angle hysteresis are important for the design of superhydrophobic materials. One more factor should be considered when superhydrophobic surfaces are created. This factor is the stability of the Cassie wetting state, which is extremely important for constituting true superhydrophobicity.[83–89] External factors such as pressure, vibrations or bouncing may promote a Cassie-Wenzel transition, accompanied by filling the surface grooves with liquid, resulting in a change in the apparent contact angle.[83–89]

When a droplet is placed on a rough surface, a diversity of metastable states is possible for a droplet corresponding to a variety of equilibrium apparent contact angles (see Figure 2.23). It should be mentioned that the "global minimum" of the free energy usually corresponds to the Cassie impregnating state discussed in Section 2.13. Passing from one metastable wetting state to another requires surmounting the energetic barrier. The origin of this barrier will be discussed in this section in detail. The design of reliefs characterized by high barriers separating the Cassie and Wenzel states is crucial for manufacturing "truly superhydrophobic", self-cleaning surfaces. Thus, the considerations supplied in this section are of highly practical importance.

First consider the time scaling of wetting transitions. Two types of wetting transitions may proceed in their relation to time scaling: rapid *adiabatic* transitions, with a fixed value of the contact angle; and *slow non-adiabatic* transitions, when a droplet

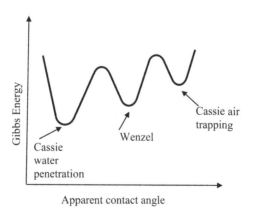

FIGURE 2.23 Sketch of multiple minima of the Gibbs energy of a droplet deposited on a rough surface. The lowest minimum corresponds to the Cassie impregnating state.

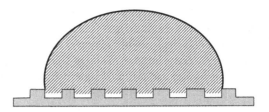

FIGURE 2.24 The composite wetting state.

has time to relax, and the contact angle changes in the course of liquid penetration into depressions (or overflowing from them).

The mechanisms of wetting transitions on inherently hydrophobic $\left(\theta_Y > \frac{\pi}{2}\right)$ vs. inherently hydrophilic surfaces are quite different and should be clearly distinguished. We start our discussion with inherently hydrophobic surfaces (the examples are polyethylene, polypropylene and polytetrafluoroethylene (Teflon)).

The Gibbs energy of a droplet deposited on a rough surface possesses multiple minima. Both the Wenzel and Cassie states occupy local minima of free energy, as shown in Figure 2.23. The Cassie air trapping wetting state usually corresponds to the highest of multiple minima of Gibbs energy of a droplet deposited on a rough hydrophobic surface (with biological and hierarchical surfaces being exceptions). Thus, for the wetting transitions, the energy barrier separating the Cassie and Wenzel states must be surmounted.[83–89] It was argued that this energy barrier corresponds to the surface energy variation between the Cassie state and a hypothetical *composite* state, with the almost complete filling of surface asperities by water, as shown in Figure 2.24, keeping the liquid-air interface under the droplet and the contact angle constant.

In contrast to the equilibrium mixed wetting state (shown in Figure 2.21 and discussed in References 63 and 64 by Miwa, Marmur et al.), the composite state is unstable for hydrophobic surfaces and corresponds to an energy maximum (transition state). For the simple topography depicted in Figure 2.25, the energy barrier could be calculated as follows (see Reference 3):

$$W_{tr} = 2\pi a^2 \frac{h}{l}(\gamma_{SL} - \gamma_{SA}) = -2\pi a^2 \frac{h}{l}\gamma\cos\theta_Y, \tag{2.32}$$

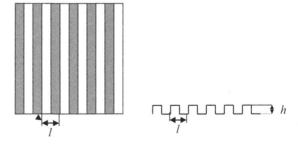

FIGURE 2.25 Geometric parameters of the model relief used for the calculation of the Cassie-Wenzel transition energy barrier.

where h and l are the geometric parameters of the relief, shown in Figure 2.25, and a is the radius of the contact area. The numerical estimation of the energetic barrier according to Eq. (2.32), with the parameters $l = h = 20\ \mu m$, $a = 1$ mm, $\theta_Y = 105°$ (corresponding to low-density polyethylene), and $\gamma = 72$ mJ·m^{-2}, gives a value of $W_{tr} = 120$ nJ. For $\theta_Y = 114°$ (corresponding to polytetrafluoroethylene, i.e., Teflon), Eq. (2.32) yields $W_{tr} = 180$ nJ. It should be stressed that according to Eq. (2.32), the energy barrier scales as $W_{tr} \sim a^2$. The validity of this assumption will be discussed below.

The energetic barrier is extremely large compared to thermal fluctuations, namely: $\frac{W_{tr}}{k_B T} \approx \left(\frac{a}{d_m}\right)^2 \gg 1$, where k_B is the Boltzmann constant and d_m is the molecular scale. At the same time, W_{tr} is much less than the energy of evaporation of the droplet: $Q \approx \frac{4}{3}\pi R^3 \lambda$, where λ is the volumetric heat of water evaporation $\left(\lambda \cong 2 \times 10^9 \frac{J}{m^3}\right)$. For a 3-$\mu l$ droplet with the radius $R \cong 1.0$ mm, it yields $Q \approx 10$ J; hence, the following hierarchy of energies takes place under wetting transitions: $k_B T \ll W_{tr} \ll Q$. Actually, this interrelation between characteristic energies is what makes wetting transitions possible.[84] Otherwise, thermal fluctuations will necessarily destroy the Cassie wetting state. If that were not the case, a droplet exposed to external stimuli might evaporate before the wetting transition.[83,89] It is instructive to estimate the radius, at which $W_{tr} \approx Q$. Equating W_{tr}, given by Eq. (2.32), to Q yields for droplets placed on PTFE (Teflon) $R \cong -(3\gamma \cos\theta_Y / 2\lambda) \cong 5 \times 10^{-11}$ m, when $R \sim a$, $h \sim l$, which is smaller than a typical molecular size $d_m \cong 3.0 - 6.0 \times 10^{-10}$ m. This means that wetting transitions are possible for any volume of a droplet. It is noteworthy that the ratio $\frac{\gamma}{\lambda}$ is practically the same for all liquids, and it is on the order of magnitude of a molecular size d_m (see the extended discussion in Sections 1.3 and 1.5 and Reference 90). Hence, wetting transitions are possible for any liquid in any volume.

It should be stressed that hierarchical reliefs, discussed in Section 2.18 and illustrated with Figure 2.22, are better at withstanding wetting transitions. Nosonovsky et al. demonstrated that curved hierarchical reliefs also provide stable equilibrium positions for the triple line.[91] Thus, it seems that preparing of hierarchical, multiscale topographies is the best recommendation for materials engineering, which are interested in the manufacturing of the self-cleaning surfaces. However, the physical reality is much more complicated and the hierarchical surfaces may demonstrate the pronounced "rose petal effect" to be discussed in detail below.[92] The rose-petal-like surfaces are strongly adhesive and do not demonstrate the self-cleaning properties.[92] The concept of hierarchical reliefs in its relation to the Gecko-effect-inspired dry adhesives will be also discussed below in Chapter 6.

It is also noteworthy that barriers separating the Cassie and Wenzel states may differ strongly for rapid (adiabatic) transitions with a fixed value of the contact angle, as opposed to slow (non-adiabatic) transitions.[84] Thus, not only thermodynamics of wetting transitions should be considered but also their kinetics.[84]

2.20   IRREVERSIBILITY OF WETTING TRANSITIONS

Wetting transitions are irreversible, i.e., spontaneous restoring of the initial wetting state is impossible. This is the main danger/problem to be considered when superhydrophobic, self-cleaning surface is developed. When liquid penetrates the details of the relief constituting superhydrophobicity, it will irreversibly remain in the grooves

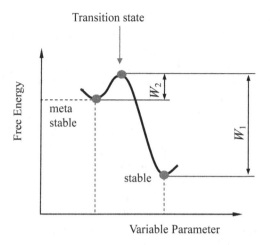

FIGURE 2.26 Sketch illustrating the irreversibility of wetting transitions. W_1 is the energetic barrier from the side of the stable state; W_2 is the energetic barrier from the side of the metastable state, $W_1 \gg W_2$. The energetic barrier is asymmetric.

and the surface will inevitably lose its useful self-cleaning properties. The concept of an energetic barrier separating the wetting states allows explanation of this irreversibility. Consider the shape of this barrier; as demonstrated in Reference 84, the energy barrier is very asymmetric, namely it is relatively low from the side of the metastable (higher-energy) state and high from the side of the stable state, as shown in Figure 2.26. Accurate calculations of energy barriers for real wetting transitions (performed in Reference 84) supplied a difference of almost one order of magnitude. Taking into account exponential (Arrhenius-type) dependence of the transition probability on the barrier height shows that a reverse transition is impossible. Remember that the arguments supplied in this section are valid for inherently hydrophobic surfaces. The above arguments are formally valid for both inherently hydrophilic and hydrophobic surfaces. However, in reality, the origin of the energetic barrier for inherently *hydrophilic* surfaces is of a different nature and will be discussed later.

2.21 CRITICAL PRESSURE NECESSARY FOR WETTING TRANSITIONS

As it always occurs, the force (pressure)-based approach to wetting transitions is possible in parallel with the energy-based one. Let us estimate the critical pressure p_c necessary for the Cassie-Wenzel wetting transition occurring on the micro-scaled surface. Consider a single-scale pillar-based biomimetic surface, similar to that studied by Yoshimitsu in Reference 93, with pillar width a, and groove width b. Analysis of the balance of forces at the air-liquid interface at which the equilibrium is still possible yielded in Reference 94:

$$p_c > \frac{\gamma f_s cos\theta_Y}{(1-f_s)\lambda}, \qquad (2.33)$$

where A and $\bar{p}$ are the pillar cross-sectional area and perimeter, respectively, and f_S is the fraction of the projection area that is wet. As an application of Eq. (2.33) with $\theta_Y = 114°$, $a = 50$ μm and $b = 100$ μm, we obtain $p_c = 296$ Pa, in an excellent agreement with experimental results.[94] Recalling that the dynamic pressure of rain droplets may be as high as 10^4–10^5 Pa, which is much larger than $p_c \cong 300$ Pa, we conclude that creating biomimetic reliefs with very high critical pressure is of practical importance.[88,89] The values of the dynamic pressure exerted by drops bouncing the rough surface may be very large due to the effect of "water hammer", as discussed in References 95 and 96.

The concept of critical pressure leads to the conclusion that reducing the microstructural scales (e.g., the pillars' diameters and spacing) is the most efficient measure to enlarge the critical pressure. Hierarchical surfaces, depicted in Figure 2.22, usually increase the value of critical pressure and the barrier separating the Cassie and Wenzel wetting states.[97–99] The dynamics of wetting transitions is discussed in References.[97–99] The characteristic time of the Cassie-Wenzel transition (i.e., the time necessary for filling microscopically scaled grooves) was established by reflection interference contrast microscopy as less than 20 ms.[98] The dynamics of wetting transitions, i.e., filling of the micro-grooves with liquid, depends on the viscosity of a liquid, introduced in Section 1.4.[100,101] Lack of experimental data related to the dynamics of wetting transitions is noteworthy. Let us demonstrate how simple qualitative considerations yield the correct order of magnitude of the time scale of wetting transitions. Assume that just viscous friction withstands filling of the details of relief (the viscosity of liquids was discussed in detail in Section 1.4). Now, we introduce kinematic viscosity of liquids υ according to Eq. (2.34):

$$\upsilon = \frac{\eta}{\rho}, \tag{2.34}$$

where η and ρ and the dynamic viscosity and density of a liquid correspondingly. Consider that the dimension of the kinematic viscosity is $[\upsilon] = \frac{m^2}{s}$. If the viscous mechanism of filling is assumed, the characteristic time of filling of the pore with the characteristic size of L is very roughly estimated according to Eq. (2.35):

$$\tau \cong \frac{L^2}{\upsilon}. \tag{2.35}$$

Assuming $L \cong 50$ μm and $\upsilon \cong 1.0 \times 10^{-6}$ $\frac{m^2}{s}$ (which is the kinematic viscosity of water at $t = 20°C$), we estimate $\tau \cong 2.5$ ms, which is close to the experimentally observed time scale of the wetting transition.[98]

2.22   THE CASSIE WETTING AND WETTING TRANSITIONS ON INHERENTLY HYDROPHILIC SURFACES

It is also noteworthy that neither the force-based or the energy-based approach explains the existence of the Cassie wetting regime on inherently hydrophilic surfaces, such as metals, observed by different groups.[80,81] Indeed, W_{tr} and p_c, calculated

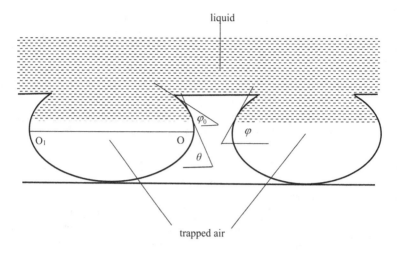

FIGURE 2.27 Geometrical air trapping on hydrophilic reliefs. The "descent" angle φ is governed by the shape of the pore; θ is the actual contact angle of the descending liquid.

according to Eqs (2.32) and (2.33), are negative for hydrophilic surfaces; this makes Cassie wetting on hydrophilic surfaces impossible.

For the explanation of the roughness-induced superhydrophobicity of inherently hydrophilic materials, it was supposed that air is entrapped by cavities constituting the topography of the surface.[102–104] The simple mechanism of "geometrical" trapping could be explained as follows: consider a hydrophilic surface (now we re-define the notion of hydrophilicity; we consider a surface as "hydrophilic" when $\theta_{adv} < \frac{\pi}{2}$ takes place) which comprises pores, embedded into the solid matrix, as depicted in Figure 2.27. It is seen that air trapping is possible only if $\theta_{adv} > \varphi_0$, where φ_0 is the angle between the tangent in the highest point of the pattern and the horizontal symmetry axis O_1O. Indeed, when the liquid level is descending, the actual angle θ is growing (see Figure 2.27), and if the condition $\theta_{adv} > \varphi_0$ is violated, the equilibrium $\theta = \theta_{adv} = \varphi$ will be impossible (recall that the advancing contact angle θ_{adv} is the equilibrium angle, although a metastable one). It should be emphasized that the phenomenon of contact angle hysteresis on flat surfaces, discussed in Section 2.7, makes the variation of θ possible.

Geometrical air trapping gives rise to an energetic barrier to be surmounted for the total filling of a pore. When water fills the hydrophilic pore (described in Figure 2.27), the energy gain due to the wetting of the pore's hydrophilic wall is overcompensated by the energy increase at the expense of the growth of the high-energetic liquid-air interface.[105] To perform the quantitative analysis, consider a spherical model of the cavity, drawn in Figure 2.28. The cavity surface energy G is expressed as follows:

$$G(\varphi) = 2\pi r^2 \gamma \cos\theta_Y \left(\cos\varphi - \cos\varphi_0\right) + \gamma\pi r^2 \sin^2\varphi, \tag{2.36}$$

where the first and second terms are the energies of the liquid-solid and liquid-air interfaces, respectively, and r is the cavity radius.[105] The energy maximum

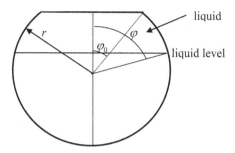

FIGURE 2.28 Formation of a transition state in a spherical cavity. Central angle φ, which defines the liquid level, is a current contact angle.

corresponds to $\varphi = \theta_Y$. Note that a central angle φ, which defines the liquid level, is, however, a current contact angle. So, the energetic barrier per cavity w from the side of the Cassie state ($\varphi = \varphi_0$) is $[w] = J$:

$$w = \pi r^2 \gamma \left(cos\varphi_o - cos\theta_Y \right)^2, \tag{2.37}$$

when the condition $\theta_Y > \varphi_o$ takes place. The counterpart of w in Eq. (2.37) per one droplet can roughly be evaluated as $W \sim \pi r^2 \gamma N$, $[W] = J$, where $N \sim \frac{S}{4r^2}$ is the number of unit cells in the liquid-solid interface area S for a plane quadratic close-packed lattice with the lattice constant $2r$.[105] Thus, for a droplet with a contact radius $a \sim 1$ mm, the upper limit $W \sim \pi S \gamma / 4 \sim 10^2$ nJ is remarkably on the same order of magnitude as the barrier inherent to microscopically scaled hydrophobic surfaces, as shown in Section 2.19. Hydrophilic surfaces of various topographies are discussed in Reference 105.

For hydrophilic materials, the Cassie state always corresponds to the higher energy state compared to the Wenzel one but is stabilized by the energy barrier. The condition for the existence of such a barrier is a geometrical property of the relief that provides the sufficient increase in the liquid-air interface in the course of the liquid penetration into details of this relief.[105] The increase of the high-energy liquid-air interface in the course of the liquid descent along the details of a relief also explains remarkable stability of re-entrant (or so called "hoodoo-like") reliefs (shown in Figure 2.29), which demonstrate not only superhydrophobicity but also oleophobicity.[106]

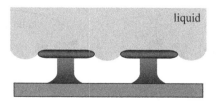

FIGURE 2.29 "Hoodoo-like" elements supplying superoleophobic properties to the surface. High-energy liquid-air interface increases abruptly under liquid descent.

The effect of line tension may increase or decrease the potential barrier separating the Cassie and the Wenzel wetting states, depending on the sign of the line tension and the topography of a relief.[107,108]

2.23 THE DIMENSION OF WETTING TRANSITIONS

We already mentioned that wetting Cassie-Wenzel transitions destroy the Cassie air trapping wetting regime responsible for the superhydrophobicity and self-cleaning of lotus-like surfaces. Thus, understanding of the physical mechanism of these transitions is crucial for the manufacturing of the self-cleaning surfaces. Generally speaking, two scenarios of wetting transitions, depicted in Figure 2.30, are possible. Not only vertical but also horizontal (lateral) de-pinning of the triple line leads to a wetting transition, as shown in Figure 2.30. Figure 2.30A depicts a vertical wetting transition under a pinned triple line, whereas Figure 2.30B demonstrates the transition under a laterally de-pinned triple line. Wetting transitions accompanied by the lateral de-pinning of the triple line were observed under vibration of droplets[109–111] and wetting transitions inspired by electrowetting.[112] The horizontal de-pinning of the triple line may lead to the so-called 1D scenario of wetting transitions. The question is: whether all pores underneath the droplet should be filled by liquid (the "2D scenario"), or perhaps only the pores adjacent to the three-phase (triple) line are filled under external stimuli such as pressure, vibrations or impact (the "1D scenario"). Indeed, the apparent contact angle is dictated by the area adjacent to the triple line and not by the total area underneath the droplet (see the discussion in Section 2.14 and References 56–61). Thus, for its change, it is sufficient to fill pores which are close to the triple line. The experiments carried out with vibrated drops and electrowetting

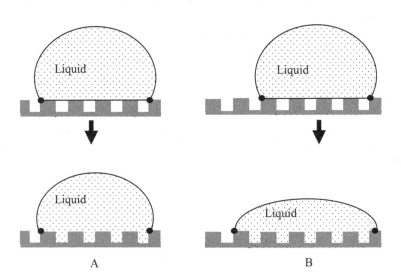

FIGURE 2.30 Scheme of two scenarios of wetting transitions. (**A**) – the triple line is pinned, (**B**) the triple line is de-pinned and displaced laterally.

supported the 1D scenario of wetting transitions.[109–112] For the detailed discussion of the problem of the dimension of wetting transitions, see Reference 97.

2.24 SUPEROLEOPHOBICITY

The design and manufacture of surfaces repelling organic oils is an important technological task. At the same time, it is an extremely challenging goal, due to the fact that organic oils possess surface tensions significantly lower than that of water (see Table 1.1, Chapter 1). Thus, typical superhydrophobic surfaces demonstrate the Wenzel "sticky" wetting when an oil drop is put onto them. Several groups succeeded in solving this problem and reported oil-repellent surfaces.[106] These surfaces comprise "hoodoo-like" elements, as depicted in Figure 2.29.[106] Aizenberg et al. proposed a witty approach to manufacturing superoleophobic surfaces, inspired by the *Nepenthes* pitcher plant, exploiting an intermediary liquid filling the grooves constituting a micro-relief in the biological tissue.[113] Well-matched solid and liquid surface energies, combined with the microtextural roughness, create a highly stable wetting state resulting in superoleophobicity.[113] Excellent review of novel methods enabling manufacturing of superoleophobic surfaces is found in Reference 114.

2.25 RECOMMENDATIONS TO MATERIALS SCIENCE ENGINEERS

If you plan to develop superhydrophobic, self-cleaning surface, consider the following points:

a. To be superhydrophobic, the surface should demonstrate: high apparent contact angle, low contact angle hysteresis and high stability of the Cassie air trapping wetting regime.
b. The main enemies of the true superhydrophobicity are the Cassie-Wenzel wetting transition, converting the surface to a high-sticky one.
c. In order to prevent the Cassie-Wenzel transitions, the critical pressure of the transition given by Eq. (2.33) should be increased. Reducing the microstructural scales (e.g., the pillars' diameters and spacing) is the most efficient measure to enlarge the critical pressure. Hierarchical surfaces increase the value of critical pressure and the barrier separating the Cassie and Wenzel wetting states.
d. Superhydrophobic surfaces are not necessarily superoleophobic. Development of re-entrant and hoodoo-like microstructures, depicted in Figure 2.29, is the most effective way to the development of superoleophobic surfaces (see Reference 114 summarizing the methods, suggested for manufacturing superoleophobic surfaces).

Bullets

- The spreading parameter $\Psi = \gamma_{SA} - (\gamma_{SL} + \gamma)$ governs the wetting regime, when $\Psi < 0$, wetting is partial, when $\Psi > 0$, wetting is complete.
- The contact angle established on the ideal surface is called the Young contact angle θ_Y, and it is given by the Young equation: $cos\theta_Y = \frac{\gamma_{SA} - \gamma_{SL}}{\gamma}$.

- Actually, the Young equation is the transversality condition for the variational problem of wetting.
- The Young contact angle is independent of the droplet shape and external fields.
- Line tension Γ arises from the unusual energetic state of molecules located at the triple line. There is currently no general agreement concerning either the value of line tension or about its sign. The contact angle is modified by the line tension according to the Neumann-Boruvka equation: $\cos\theta = \frac{\gamma_{SA}-\gamma_{SL}}{\gamma} - \frac{\Gamma}{\gamma a}$.
- Wetting of very thin liquid layers is governed to a large extent by disjoining pressure.
- Droplets with characteristic dimensions much less than $l_{ca} = \sqrt{\frac{\gamma}{\rho g}}$ (the capillary length) keep their spherical shape; larger drops are distorted by gravity.
- A spectrum of contact angles is possible for a certain solid/liquid pair. Maximal and minimal contact angles are called advancing and receding contact angles. The phenomenon is called the contact angle hysteresis. The contact angle hysteresis is observed even on ideal, atomically flat substrates due to the pinning of the contact (triple) line. Contact angle hysteresis is strengthened by the roughness and chemical heterogeneity of a substrate.
- When triple line moves, wetting is characterized by the dynamic contact angle which is different from the Young angle. The dynamic contact angle is given by the Cox-Voinov law. An interplay between viscosity and surface tension-related effects is described by the capillary number $Ca = \frac{\eta \tilde{v}}{\gamma}$.
- Wetting of rough or chemically heterogeneous surfaces is described by the apparent contact angle which may be introduced when the characteristic size of a droplet is much larger than that of the surface heterogeneity or roughness.
- Wetting of rough chemically homogeneous surfaces is described by the Wenzel equation. Surface roughness always magnifies the underlying wetting properties.
- Wetting of flat chemically heterogeneous surfaces is described by the Cassie-Baxter equation. The Cassie-Baxter model may be extended to a situation where a droplet traps air, i.e., it is supported partially by a solid and partially by air. One more wetting regime is possible, i.e., the Cassie-Baxter impregnating state when a drop is deposited on a substrate which comprises a patchwork of solid and liquid, where solid "islands" ahead of the drop are dry.
- The mixed wetting regime corresponds to the situation where a droplet is supported by a rough solid surface and air.
- The area adjacent to the triple line is of primary importance for predicting apparent contact angles.
- The apparent contact angles, predicted by the Wenzel and Cassie-Baxter models, are independent of external fields, volume and shape of droplets.

EXERCISES

1. Liquid is placed on the ideal solid surface. The triad of the interfacial tensions is $\gamma_{SA} = 200\,\frac{mJ}{m^2}$; $\gamma_{SL} = 22\,\frac{mJ}{m^2}$; $\gamma = 71\,\frac{mJ}{m^2}$. What may be concluded about the wetting regime inherent for this system?

   Answer: The complete (total) wetting is expected.

2. Liquid is placed on the ideal solid surface. The spreading parameter is negative.
   $\gamma_{SA} = 22 \frac{mJ}{m^2}$; $\gamma_{SL} = 30 \frac{mJ}{m^2}$. What may be concluded about the wetting regime inherent for this system?
   Answer: The partial wetting will be observed. The contact angle will be obtuse.

3. Explain qualitatively phenomenon of linear tension. What are the dimensions of the linear tension? What are the typical values of the linear tension?

4. Explain qualitatively the origin of the disjoining pressure.

5. Calculate the Wenzel contact angle for the Teflon surface ($\theta_Y = 114^0$), when $\tilde{r} = 1.5$.

6. Calculate the critical pressure corresponding to the Cassie-Wenzel wetting transition for the relief for a single-scale pillar based biomimetic surface built of Teflon pillars ($\theta_Y = 114^0$), with width a, and groove width b, $a = 25$ μm and $b = 50$ μm.
   Hint: Use Eq. (2.33).

7. Calculate the energy barrier of wetting transition for water droplets placed on the Teflon surface ($\theta_Y = 114^0$), shown in Figure 2.25, is the radius of the contact area $a = 0.5$ mm. The geometrical parameters of the relief $l = h = 50$ μm.
   Hint: Use Eq. (2.32).

8. Why wetting transitions are irreversible?

REFERENCES

1. Young T. An essay on the cohesion of liquids. *Phil. Trans. R. Soc. London* 1805, **95**, 65–87.
2. de Gennes P. G., Brochard-Wyart F., Quéré D. *Capillarity and Wetting Phenomena*, Springer, Berlin, 2003.
3. Bormashenko E., *Wetting of Real Surfaces*, 2nd Ed., de Gruyter, Berlin, 2017.
4. Bormashenko E. Young, Boruvka-Neumann, Wenzel and Cassie-Baxter equations as the transversality conditions for the variational problem of wetting. *Colloids Surf. A* 2009, **345**, 163–165.
5. Amirfazli A., Neumann A. W. Status of the three-phase line tension. *Adv. Colloid Interface Sci.* 2004, **110**, 121–141.
6. Checco A., Guenoun P. Nonlinear dependence of the contact angle of nanodroplets on contact line curvature. *Phys. Rev. Lett.* 2003, **91** (18), 186101.
7. Bormashenko E. Entropy contribution to the line tension: Insights from polymer physics, water string theory, and the three-phase tension. *Entropy* 2018, **20** (9), 712.
8. Rowlinson J. S., Widom B. *Molecular Theory of Capillarity*, Clarendon, Oxford, UK, 1982.
9. Rubinstein M., Colby R. H. *Polymer Physics*, Oxford University Press, Oxford, UK, 2003.
10. Pompe T., Fery A., Herminghaus S. Measurement of Contact Line Tension by Analysis of the Three-Phase Boundary with Nanometer Resolution, in Apparent and Microscopic Contact Angles; Drelich J., Laskowski J. S., Mittal K. L., Eds.; VSP, Utrecht, 2000; pp. 3–12.
11. Marmur A. Line tension and the intrinsic contact angle in solid-liquid-fluid systems. *J. Colloid Interface Sci.* 1997, **186**, 462–466.
12. Irajizad P., Nazifi S., Ghasemi H. Icephobic surfaces: Definition and figures of merit. *Adv. Colloid Interface Sci.* 2019, **269**, 203–218.
13. Israelachvili J. N. *Intermolecular and Surface Forces*, 3rd Ed., Elsevier, Amsterdam, 2011.

14. Erbil H. Y. *Surface Chemistry of Solid and Liquid Interfaces*, Blackwell, Oxford, 2006.
15. Derjaguin B. V., Churaev N. V. Structural component of disjoining pressure. *J. Colloid Interface Sci.* 1974, **49**, 249–255.
16. Derjaguin B. V., Churaev N. V. On the question of determining the concept of disjoining pressure and its role in the equilibrium and flow of thin films. *J. Colloid Interface Sci.* 1978, **66** (3), 389–398.
17. Nepomnyashchy A. A., Simanovskii I. B. Marangoni instability in ultrathin two-layer films. *Phys. Fluids* 2007, **19**, 122103.
18. Dai B., Leal L. G., Redondo A. Disjoining pressure for nonuniform thin films. *Phys. Rev. E* 2008, **78**, 061602.
19. Starov V. M., Velarde M. G. Surface forces and wetting phenomena. *J. Phys: Condens. Matter* 2009, **21**, 464121.
20. Hejazi V., Moghadam A. D., Rohatgi P., Nosonovsky M. Beyond Wenzel and Cassie–Baxter: Second-order effects on the wetting of rough surfaces. *Langmuir* 2014, **30** (31), 9423–9429.
21. Bormashenko E., Starov V. Impact of surface forces on wetting of hierarchical surfaces and contact angle hysteresis. *Colloid. Polym. Sci.* 2013, **291**, 343–346.
22. Zisman G. A., Todes O. M. Course of General Physics, 6th Ed. Dnipro, Kiev, Ukraine, 1994; Vol. 1 (in Russian).
23. Eloy C. Leonardo's rule, self-similarity, and wind-induced stresses in trees. *Phys. Rev. Lett.* 2011, **107**, 258101.
24. Nosonovsky M., Roy P. Allometric scaling law and ergodicity breaking in the vascular system. *Microfluid Nanofluid* 2020, **24**, 53.
25. Brown H. R. The theory of the rise of sap in trees: Some historical and conceptual remarks. *Phys. Perspect.* 2013, **15**, 320–358.
26. Kohonen M. M. Engineered wettability in tree capillaries. *Langmuir* 2006, **22**, 3148–3153.
27. Zimmermann U., Schneider H., Wegner L. H., Haase A. Water ascent in tall trees: Does evolution of land plants rely on a highly metastable state. *New Phytologist* 2004, **162**, 575–615.
28. Scholander P. F., Hammel H. T., Bradstreet E. D., Hemmingsen E. A. Pressure in vascular plants. *Science* 1965, **148**, 339–346.
29. Tadmor R. Open problems in wetting phenomena: Pinning retention forces. *Langmuir* 2021, **37** (21), 6357–6372.
30. Tadmor R., Yadav P. S. As-placed contact angles for sessile drops. *J. Colloid Interface Sci.* 2008, **317**, 241–246.
31. Tadmor R. Approaches in wetting phenomena. *Soft Matter* 2011, **7**, 1577–1580.
32. Bormashenko E. Wetting of real solid surfaces: New glance on well-known problems. *Colloid Polym. Sci.* 2013, **291**, 339–342.
33. Extrand C. W., Kumagai Y. An experimental study of contact angle hysteresis. *J. Colloid Interface Sci.* 1997, **191**, 378–383.
34. Lam C. N. C., Wu R., Li D., Hair M. L., Neumann A. W. Study of the advancing and receding contact angles: Liquid sorption as a cause of contact angle hysteresis. *Adv. Colloid Interface Sci.* 2002, **96**, 169–191.
35. Yaminsky V. V. Hydrophobic Transitions, in *Apparent and Microscopic Contact Angles*; Drelich J., Laskowski J. S., Mittal K. L., Eds.; VSP, Utrecht, 2000; pp. 47–93.
36. Butt H. J., Berger R., Steffen W., Vollmer D., Weber S. Adaptive wetting—Adaptation in wetting. *Langmuir* 2018, **34** (38), 11292–11304.
37. Marmur A., Della Volpe C., Siboni S., Amirfazli A., Drelich J. W. Contact angles and wettability: Towards common and accurate terminology. *Surface Innovations* 2017, **5** (1), 3–8.
38. Iliev S., Pesheva N., Iliev P. Contact angle hysteresis on doubly periodic smooth rough surfaces in Wenzel's regime: The role of the contact line depinning mechanism. *Phys. Rev. E* 2018, **97**, 042801.

39. Marmur A. Contact angle hysteresis on heterogeneous smooth surfaces. *J. Colloid Interface Sci.* 1994, **168**, 40–46.
40. Voinov O. V. Hydrodynamics of wetting. *Fluid Dynamics* 1976, **11**, 714–721.
41. Hoffman R. L. A study of the advancing interface. *J. Colloid Interface Sci.* 1975, **50**, 228–241.
42. Bonn D., Eggers J., Indekeu J., Meunier J., Rolley E. *Rev. Mod. Phys.* 2009, **81**, 739–805.
43. Pahlavan A. A., Cueto-Felgueroso L., McKinley G. H., Juanes R. Thin films in partial wetting: Internal selection of contact-line dynamics. *Phys. Rev. Lett.* 2015, **115**, 034502.
44. Cassie A. B. D., Baxter S. Wettability of porous surfaces. *Trans. Faraday Soc.* 1944, **40**, 546–551.
45. Cassie A. B. D. Contact angles. *Discuss. Faraday Soc.* 1948, **3**, 11–16.
46. Milne A. J. B., Amirfazli A. The Cassie equation: How it is meant to be used. *Adv. Colloid Interface Sci.* 2012, **170**, 48–55.
47. Bhushan B., *Biomimetics. Bioinspired Hierarchical-Structured Surfaces for Green Science and Technology.* 3rd Ed., Springer Nature, Cham, Switzerland, 2018.
48. Bormashenko E. General equation describing wetting of rough surfaces. *J. Colloid Interface Sci.* 2011, **360**, 317–319.
49. Bormashenko E. Why does the Cassie–Baxter equation apply? *Colloids Surf. A* 2008, **324**, 47–50.
50. Bico J., Thiele U., Quéré D. Wetting of textured surfaces. *Colloids Surf. A* 2002, **206**, 41–46.
51. Bormashenko E., Pogreb R., Stein T., Whyman G., Erlich M., Musin A., Machavariani V., Aurbach D. Characterization of rough surfaces with vibrated drops. *Phys. Chem. Chem. Phys.* 2008, **27**, 4056–4061.
52. Huang W., Lei M., Huang H., Chen J., Chen H. Effect of polyethylene glycol on hydrophilic TiO2 films: Porosity-driven superhydrophilicity. *Surf. Coatings Technol.* 2010, **204** (24), 3954–3961.
53. Nam K., Abdulhafez M., Tomaraei N. G., Bedewy M. Laser-induced fluorinated graphene for superhydrophobic surfaces with anisotropic wetting and switchable adhesion. *Appl. Surf. Sci.* 2022, **574**, 151339.
54. Gao L., McCarthy T. J. How Wenzel and Cassie were wrong? *Langmuir* 2007, **23**, 3762–3765.
55. Gao L., McCarthy T. J. An attempt to correct the faulty intuition perpetuated by the Wenzel and Cassie "Laws". *Langmuir* 2009, **25**, 7249–7255.
56. McHale G. Cassie and Wenzel: Were they really so wrong? *Langmuir* 2007, **23**, 8200–8205.
57. Panchagnula M. V., Vedantam S. Comment on how Wenzel and Cassie were wrong. *Langmuir* 2007, **23**, 13242.
58. Nosonovsky M. On the range of applicability of the Wenzel and Cassie equations. *Langmuir* 2007, **23**, 9919–9920.
59. Marmur A. When Wenzel and Cassie are right: Reconciling local and global considerations. *Langmuir* 2009, **25**, 1277–1281.
60. Bormashenko E. A variational approach to wetting of composite surfaces: Is wetting of composite surfaces a one-dimensional or two-dimensional phenomenon? *Langmuir* 2009, **25**, 10451–10454.
61. Bormashenko E., Bormashenko Y. Wetting of composite surfaces: When and why is the area far from the triple line important? *J. Phys. Chem. C* 2013, **117** (38), 19552–19557.
62. Sun C., McClure J., Berg S., Mostaghimi P., Armstrong R. T. Universal description of wetting on multiscale surfaces using integral geometry. *J. Colloid Interface Sci.* 2022, **608** (3), 2330–2338.
63. Miwa M., Nakajima A., Fujishima A., Hashimoto K., Watanabe T. Effects of the surface roughness on sliding angles of water droplets on superhydrophobic surfaces. *Langmuir* 2000, **16**, 5754–5760.

64. Marmur A. Wetting on hydrophobic rough surfaces: To be heterogeneous or not to be? *Langmuir* 2003, **19**, 8343–8348.
65. Wong T.-S., Ho C. M. Dependence of macroscopic wetting on nanoscopic surface textures. *Langmuir* 2009, **25**, 12851–12854.
66. Bormashenko E. Variational framework for defining contact angles: A general thermodynamic approach. *J. Adhesion Sci. & Technology* 2020, **34**(2), 219–230.
67. Barthlott W., Neinhuis C. Purity of the sacred lotus, or escape from contamination in biological surfaces. *Planta* 1997, **202**, 1–8.
68. Koch K., Bhushan B., Barthlott W. Multifunctional surface structures of plants: An inspiration for biomimetics. *Prog. Mater. Sci.* 2009, **54**, 137–178.
69. Yan Y. Y., Gao N., Barthlott W. Mimicking natural superhydrophobic surfaces and grasping the wetting process: A review on recent progress in preparing superhydrophobic surfaces. *Adv. Colloid Interface Sci.* 2011, **169**, 80–105.
70. Barthlot W., Mail M., Neinhuis C. Superhydrophobic hierarchically structured surfaces in biology: Evolution, structural principles and biomimetic applications. *Phil. Trans. Royal Soc. A* 2016, **374** (2073), 1–41.
71. Barthlott W., Mail M., Bhushan B., Koch K. *Plant Surfaces: Structures and Functions for Biomimetic Applications*; Bhushan, B., Ed.; Springer Handbook of Nanotechnology. Springer Handbooks. Springer, Berlin, Heidelberg, 2017.
72. Khan M. Z., Militky J., Petru M., Tomkováa B., Ali A., Tören E., Perveen S. Recent advances in superhydrophobic surfaces for practical applications: A review. *Eur. Polymer J.* 2022, **178**, 111481.
73. Guo Z., Liu W. Biomimic from the superhydrophobic plant leaves in nature: Binary structure and unitary structure. *Plant Sci.* 2007, **172** (6), 1103–1112.
74. Zheng L., Cao C., Cao L., Chen Z., Huang Q., Song B. Bounce behavior and regulation of pesticide solution droplets on rice leaf surfaces. *J. Agric. Food Chem.* 2018, **66** (44), 11560–11568.
75. Zheng Y., Gao X., Jiang L. Directional adhesion of superhydrophobic butterfly wings. *Soft Matter* 2007, **3**, 178–182.
76. Feng X. Q., Gao X., Wu Z., Jiang L., Zheng Q. S. Superior water repellency of water strider legs with hierarchical structures: Experiments and analysis. *Langmuir* 2007, **23**, 4892–4896.
77. Yin W., Zheng Y. L., Lu H. Y., Zhang X. J., Tian Y. Three-dimensional topographies of water surface dimples formed by superhydrophobic water strider legs. *Appl. Phys. Lett.* 2016, **109**, 163701.
78. Bormashenko E., Bormashenko Y., Stein T., Whyman G., Bormashenko E. Why do pigeon feathers repel water? Hydrophobicity of pennae, Cassie–Baxter wetting hypothesis and Cassie–Wenzel capillarity-induced wetting transition. *J. Colloid Interface Sci.* 2007, **311** (1), 212–216.
79. Herminghaus S. Roughness-induced non wetting. *Europhys. Lett.* 2000, **52**, 165–170.
80. Ellinas K., Dimitrakellis P., Sarkiris P., Gogolides E. A review of fabrication methods, properties and applications of superhydrophobic metals. *Processes* 2021, **9** (4), 666.
81. Wang N., Xiong D. Superhydrophobic membranes on metal substrate and their corrosion protection in different corrosive media. *Appl. Surf. Sci.* 2014, **305**, 603–608.
82. Grynyov R., Bormashenko E., Whyman G., Bormashenko Y., Musin A., Pogreb R., Starostin A., Valtsifer V., Strelnikov V., Schechter A., Kolagatla S. Superoleophobic surfaces obtained via hierarchical metallic meshes. *Langmuir* 2016, **32** (17), 4134–4140.
83. Ishino C., Okumura K., Quéré D. Wetting transitions on rough surfaces. *Europhys. Lett.* 2004, **68**, 419–425.
84. Whyman G., Bormashenko E. Wetting transitions on rough substrates: General considerations. *J. Adhes. Sci. Technol.* 2012, **26**, 2012.

85. Patankar N. A. On the modeling of hydrophobic contact angles on rough surfaces. *Langmuir* 2003, **29**, 1249–1253.
86. Patankar N. A. Transition between superhydrophobic states on rough surfaces. *Langmuir* 2004, **20**, 7097–7102.
87. Barbieri L., Wagner E., Hoffmann P. Water wetting transition parameters of perfluorinated substrates with periodically distributed flat-top microscale obstacles. *Langmuir* 2007, **23**, 1723–1734.
88. Wang X., Fu C., Zhang C., Qiu Z., Wang B. A comprehensive review of wetting transition mechanism on the surfaces of microstructures from theory and testing methods. *Materials* 2022, **15** (14), 4747.
89. He X., Zhang B. X., Wang S. L., Wang Y. F., Yang Y. R., Wang X. D., Lee D. J. The Cassie-to-Wenzel wetting transition of water films on textured surfaces with different topologies. *Phys. Fluids* 2021, **33**, 112006.
90. Bormashenko E. Why are the values of the surface tension of most organic liquids similar? *Am. J. Phys.* 2010, **78**, 1309–1311.
91. Nosonovsky M., Bhushan B. Hierarchical roughness makes superhydrophobic states stable. *Microelectron. Eng.* 2007, **84**, 382–386.
92. Fernández A., Francone A., Thamdrup L. H., Johansson A., Bilenberg B., Nielsen T., Guttmann M., Sotomayor Torres C. M., Kehagias N. Design of hierarchical surfaces for tuning wetting characteristics. *ACS Appl. Mater. Interfaces* 2017, **9** (8), 7701–7709.
93. Yoshimitsu Z., Nakajima A., Watanabe T., Hashimoto K. Effects of surface structure on the hydrophobicity and sliding behavior of water droplets. *Langmuir* 2002, **18**, 5818–5822.
94. Zheng Q. S., Yu Y., Zhao Z. Effects of hydraulic pressure on the stability and transition of wetting modes of superhydrophobic surfaces. *Langmuir* 2005, **21**, 12207–12212.
95. Sarker S., Sarker T. Spectral properties of water hammer wave. *Appl. Mech.* 2022, **3** (3), 799–814.
96. Fujisawa K. Effect of impact velocity on time-dependent force and droplet pressure in high-speed liquid droplet impingement. *Ann. Nucl. Energy* 2022, **166**, 108814.
97. Bormashenko E. Progress in understanding wetting transitions on rough surfaces. *Adv. Colloid Interface Sci.* 2015, **222**, 92–103.
98. Moulinet S., Bartolo D. Life and death of a fakir droplet: Impalement transitions on superhydrophobic surfaces. *Eur. Phys. J. E* 2007, **24**, 251–260.
99. Papadopoulos P., Mammen L., Deng X., Vollmer D., Butt H. J. How superhydrophobicity breaks down. *Proc. Natl. Acad. Sci. USA* 2013, **110**, 3254–3258.
100. Mao Y., Chen Y., Xia W., Peng Y., Xie G. Understanding the dynamic pore wetting by 1H LF-NMR characterization. Part 1: Effect of dynamic viscosity of liquid. *Colloids Surf. A* 2021, **613**, 126039.
101. Pan B., Clarkson C. R., Atwa A., Debuhr C., Ghanizadeh A., Birss V. I. Wetting dynamics of nanoliter water droplets in nanoporous media. *J. Colloid Interface Sci.* 2021, **589**, 411–423.
102. Bormashenko E., Bormashenko Y., Whyman G., Pogreb R., Stanevsky O. Micrometrically scaled textured metallic hydrophobic interfaces validate the Cassie–Baxter wetting hypothesis. *J. Colloid Interface Sci.* 2006, **302**, 308–311.
103. Patankar N. A. Hydrophobicity of surfaces with cavities: Making hydrophobic substrates from hydrophilic materials? *J. Adhes. Sci. Technol.* 2009, **23**, 413–433.
104. Wang J., Chen D. Criteria for entrapped gas under a drop on an ultrahydrophobic surface. *Langmuir* 2008, **24**, 10174–10180.
105. Whyman G., Bormashenko E. How to make the Cassie wetting state stable? *Langmuir* 2011, **27**, 8171–8176.
106. Tuteja A., Choi W., Mabry J. M., McKinley G. H., Cohen R. E. Robust omniphobic surfaces. *Proc. Natl. Acad. Sci. USA* 2008, **105**, 18200–18205.

107. Bormashenko E., Whyman G. On the role of the line tension in the stability of Cassie wetting. *Langmuir* 2013, **29**, 5515–5519.

108. Iwamatsu M. Free-energy barrier of filling a spherical cavity in the presence of line tension: Implication to the energy barrier between the Cassie and Wenzel states on a superhydrophobic surface with spherical cavities. *Langmuir* 2016, **32** (37), 9475–9483.

109. Bormashenko E., Pogreb R., Whyman G., Erlich M. Cassie-Wenzel wetting transition in vibrating drops deposited on rough surfaces: Is dynamic Cassie-Wenzel wetting transition a 2D or 1D affair? *Langmuir* 2007, **23**, 6501–6503.

110. Bormashenko E., Pogreb R., Whyman G., Erlich M. Resonance Cassie-Wenzel wetting transition for horizontally vibrated drops deposited on a rough surface. *Langmuir* 2007, **23**, 12217–12221.

111. Sudeepthi A., Yeo L., Sen A. K. Cassie–Wenzel wetting transition on nanostructured superhydrophobic surfaces induced by surface acoustic waves. *Appl. Phys. Lett.* 2020, **116**, 093704.

112. Bahadur V., Garimella S. V. 1D preventing the Cassie-Wenzel transition using surfaces with noncommunicating roughness elements. *Langmuir* 2009, **25**, 4815–4820.

113. Wong T. S., Kang S. H., Tang S. K. Y., Smythe T. J., Hatton B. D., Grinthal A., Aizenberg J. Bioinspired self-repairing slippery surfaces with pressure-stable omniphobicity. *Nature* 2011, **477**, 443–447.

114. Liu K., Tian Y., Jiang L. Bio-inspired superoleophobic and smart materials: Design, fabrication, and application. *Progr. Mater. Sci.* 2013, **58** (4), 503–564.

3 The Rose Petal Effect

It was already mentioned in Section 2.25 that high apparent contact angles are necessary but not sufficient for true superhydrophobicity accompanied by self-cleaning properties of a surface. Jiang et al. reported that rose petal surfaces demonstrate high contact angles attended with extremely high contact angles hysteresis.[1] The surface of the rose petal is built from hierarchically riffled "micro-bumps" resembling those of lotus leaves.[1] At the same time, the wetting of rose petals is very different from that of lotus leaves. The apparent angles of droplets placed on a rose petal are high, but the droplets are simultaneously in a "sticky" wetting state; they do not roll.[1] Jiang called this phenomenon the "rose petal effect".[1,2] Water droplet placed on the rose petal is depicted in Figure 3.1. The surfaces made of lycopodium particles also demonstrate the pronounced rose petal effect as reported in Reference 2 and illustrated in Figure 3.2.

FIGURE 3.1 Water droplet placed on the rose petal. The apparent contact angle is high. However, the droplet does not roll from the rose petal. Image made in the laboratory of interface science of the Ariel University.

FIGURE 3.2 A 10-μl droplet deposited on a surface built of lycopodium particles (for details, see Reference 2).

DOI: 10.1201/9781003178477-3   **65**

The natural explanation for the "rose petal effect" is supplied by the Wenzel model (see Section 2.10). Inherently hydrophobic flat surfaces may demonstrate apparent contact angles approaching π when rough (see Eq. (2.24)). Wenzel wetting is characterized by high contact angle hysteresis; thus, the experimental situation depicted in Figure 3.2, when droplet does not slide down the surface, when the surface is vertical, becomes possible.

However, the Wenzel model does not explain the existence of the "rose petal effect" for inherently hydrophilic surfaces; indeed, inherently hydrophilic surfaces become more hydrophilic in the Wenzel wetting regime (see Section 2.10). Bhushan and Nosonovsky demonstrated that wetting of hierarchical reliefs may be of a complicated nature, resulting in the "rose petal effect", as shown in Figure 3.3 and discussed in Reference 3. The detailed microscopic study of the rose petals was reported in Reference 4. Various wetting modes are possible for hierarchical reliefs: it is possible that a liquid fills the larger grooves, whereas small-scale grooves are not wetted and trap air as shown in Figure 3.3A. The inverse situation is also possible in which small-scale grooves are wetted and large-scale ones form air cushions (see Figure 3.3B). According to Reference 3, the larger structure controls the contact angle hysteresis, whereas the smaller (usually nanometric) scale is responsible for high contact angles.[1-3] Thus, the relief depicted in Figure 3.3A will demonstrate high contact angles attended by high contact angle hysteresis. This hypothesis reasonably explains the "rose petal effect". However, it is clearly seen that a broad variety of wetting modes is possible on hierarchical surfaces, opening the way to a diversity of technological applications of hierarchically rough surfaces. Shape memory polymers enabled development of the surfaces which display a multiply switchable transition between the "lotus-effect" and the "rose-petal effect", revealing a potential for rewritable liquid patterns, controlled droplet transportation.[5]

The alternative model of the rose petal effect was suggested in Reference 6. Apparent contact angles are totally governed by the area of the solid surface adjacent to the triple (three-phase) line.[6] However, the energy of adhesion, in turn, depends on the physical and chemical properties of the entire area underneath the droplet.[6] Thus, a droplet deposited axisymmetrically on the superhydrophobic surface comprising a non-superhydrophobic spot holds "sticky" wetting attended with high apparent contact angles, thus demonstrating the rose petal effect.[6]

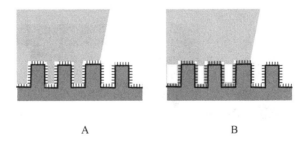

A   B

FIGURE 3.3   Scheme of various wetting scenarios possible on a hierarchical relief.[3]

Bullets

- Surfaces demonstrating simultaneously the high apparent contact angles and high adhesion were reported.
- High apparent contact angles accompanied with high adhesion are inherent for rose petals.
- The rose petal effect is reasonably attributed to the wetting of hierarchical surfaces.
- Artificial and switchable rose-petal-like surfaces were reported.

EXERCISES

3.1. Explain how wetting of hierarchical surfaces gives rise to the rose petal effect.

REFERENCES

1. Feng L., Zhang Y., Xi J., Zhu Y., Wang N., Xia F., Jiang L. Petal effect: A superhydrophobic state with high adhesive force. *Langmuir* 2008, **24**, 4114–4119.
2. Bormashenko E., Stein T., Pogreb R., Aurbach D. "Petal effect" on surfaces based on lycopodium: High stick surfaces demonstrating high apparent contact angles. *J. Phys. Chem. C* 2009, **113**, 5568–5572.
3. Bhushan B., Nosonovsky M. The rose petal effect and the modes of superhydrophobicity. *Philos. Trans. Royal Soc. A* 2010, **368**, 4713–4728.
4. Almonte L., Pimentel C., Rodríguez-Cañas E., Abad J., Fernández V., Colchero J. Rose petal effect: A subtle combination of nano-scale roughness and chemical variability. *Nanoselect* 2022, **3** (5), 977–989.
5. Shao Y., Zhao J., Fan Y., Zhang Z., Ren L. Shape memory superhydrophobic surface with switchable transition between "lotus effect" to "rose petal effect". *Chem. Eng. J* 2020, **382**, 122989.
6. Bormashenko E., Bormashenko Y. Wetting of composite surfaces: When and why is the area far from the triple line important? *J. Phys. Chem. C* 2013, **117** (38), 19552–19557.

4 Salvinia Effect

Underwater superhydrophobicity by floating fern *Salvinia* and backswimmer *Notonecta* is discussed. Leaves of the floating *Salvinia* fest demonstrate remarkable stability of the Cassie air-trapping wetting state. This stability was related to the high energy necessary for disconnection of the three-phase line inherent for these surfaces. Numerous technologies enabling mimicking of the Salvinia surfaces were reported. *Salvinia*-inspired surfaces enable drag reduction.

4.1 UNDERWATER SUPERHYDROPHOBICITY

Numerous technological applications demand materials, which remain superhydrophobic under water. In other words, we need a material, which will provide the Cassie air-trapping wetting state, introduced into Section 2.12 in underwater conditions. This task is extremely challenging due to the Cassie-Wenzel wetting transitions, discussed in Sections 2.19–2.23. The nature-inspired solution to the problem is supplied by the floating fern *Salvinia*[1] and backswimmer *Notonecta*.[2]

The unique heterogeneous eggbeater structures, depicted schematically in Figure 4.1, endow *Salvinia* leaves with superhydrophobicity and strong adhesion, which ensures that the leaves show durable air-retainability in underwater environments.[3,4]

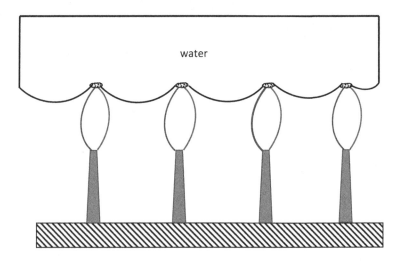

FIGURE 4.1 A leaf of *Salvinia molesta* with a water droplet is depicted; the egg-beater hairs are shown schematically, as well as the elaborate margin for the edge effect to prevent the escape of the air layer under water.

  DOI: 10.1201/9781003178477-4

Superhydrophobicity of *Salvinia* leaves demonstrates unique thermodynamic stability, preventing the Cassie-Wenzel wetting transitions. What is the physical reasoning for this stability? The answer to this question was suggested in Reference 5 and it was related to the effect of de-pinning of the triple (three-phase) line. We discussed thermal stability of the Cassie wetting in Section 2.19, where the energetic barrier separating the Cassie and Wenzel wetting states was calculated (see Eq. (2.32)). However, there exists one more factor, promoting the stability of the Cassie wetting and this is pinning of the triple line. We already addressed this effect in Section 2.7 when we spoke about the effect of the contact angle hysteresis. Pinning of the triple line also increases the barrier separating the Cassie and Wenzel wetting states. Indeed, the resulting energetic barrier to be surmounted for the Cassie-Wenzel transitions equals

$$U = U_1 + U_2, \tag{4.1}$$

where U_1 is the potential barrier arising from the filling of hydrophobic pores, already calculated in Sections 2.19 and 2.20, and U_2 is the barrier to be surmounted under disconnection (de-pinning) of the triple line. The units of both U_1 and U_2 are joules and they are related to the entire droplet, when $U_1 = S\tilde{U}_1$ and $U_2 = l\tilde{U}_2$, 2, where S and l are the area underneath the droplet filled by water and the perimeter of the triple line, respectively, $[\tilde{U}_1] = \frac{J}{m^2}$ is the specific energy of the pores filling which is the 2D effect and $[\tilde{U}_2] = \frac{J}{m}$ is the specific energy of the de-pinning of the triple line which is the 1D effect.[5,6] Calculation of $\tilde{U}_1$ was performed in Section 2.19; calculation of $\tilde{U}_2$ was carried out in Reference 7 and its value was experimentally established in Reference 8 for a diversity of materials. It was demonstrated that the value of this barrier was $\tilde{U}_2 \sim 10^{-6} \frac{J}{m}$ when water droplet was placed on various polymer substrates.[8] It was shown in Reference 6 that when only the nearest to the triple-line pores are filled in the course of the wetting transition (this scenario of the Cassie-Wenzel transition was called the 1D transition in Section 2.23 and illustrated with Figure 2.30), the energy of filling pores U_1 and the energy of the triple-line de-pinning U_2 are comparable at a quite reasonable posts' height $h \sim 10 \ \mu m$, which is typical for superhydrophobic surfaces.[6] The unusual stability of the Cassie air-trapping wetting state was related to the high energy necessary for de-pinning of the triple line on these surfaces.[5] It is noteworthy that the exhausting explanation of the *Salvinia* effect is not obtained yet. The methods enabling manufacturing artificial surfaces mimicking those resembling the *Salvinia molesta* ones are reported in References 4 and 9. Lithography, low-temperature chemical vapor deposition, atmosphere pressure plasma chemical vapor deposition, water-assisted chemical vapor deposition, electrodeposition, electrospinning, electrostatic, flocking, 3D printing, plasma etching and chemical etching were used for mimicking of egg-beater-like surfaces schematically depicted in Figure 4.1.[4] *Salvinia*-inspired surfaces were also used as drag-reducing coatings (up to 30% reduction were previously measured on the first prototypes.[3,10] We'll discuss the effect of the drag reduction when we'll address the shark skin effect.

Bullets

- *Salvinia molesta* fern leaves demonstrate remarkable stability of the Cassie-Baxter air-trapping wetting state.
- This stability was related by the researchers to the high pinning of the triple line by the egg-beater elements constituting the surface.
- Energetic barrier separating the Cassie and Wenzel wetting states emerging from the energy need for filling hydrophobic pores may be comparable to that emerging from the pinning of the triple line.

EXERCISES

4.1. A droplet on a flat substrate takes the shape of a spherical cap, with the contact radius R_0 and the equilibrium contact angle θ_0. In the course of the droplet evaporation, we observe the stick–slip motion of the triple line (see References 7 and 8). During evaporation, the droplet may be in the state with a larger contact radius than the equilibrium state (for the same volume of the droplet), $R = R_0 + \delta R$, and with a lower contact angle, $\theta = \theta_0 - \delta\theta$. Calculate the free energy of the droplet within the approach, developed by M. Shanahan in Reference 7, for the water droplet $\left(\gamma = 72.0 \frac{mJ}{m^2}\right)$, with $R = 1\,mm$, $\theta_0 = 90°$.
Solution: The free energy of the droplet is calculated as follows (see References 7 and 8):

$$G(R,\theta) = \pi\gamma R^2 \left[\frac{2}{1+cos\theta} - cos\theta_0\right].$$

Substitution of the numerical values of the parameters yields: $G \cong 3.9 \times 10^{-7}\,J$.

REFERENCES

1. Barthlott W., Wiersch S., Čolić Z., Koch K. Classification of trichome types within species of the water fern Salvinia, and ontogeny of the egg-beater trichomes. *Botany* 2009, **87** (9), 830–836.
2. Ditsche-Kuru P., Schneider E. S., Melskotte J. E., Brede M., Leder A., Barthlott W. Superhydrophobic surfaces of the water bug *Notonecta glauca*: A model for friction reduction and air retention. *Beilstein J. Nanotechnol* 2011, **2**, 137–144.
3. Barthlott W., Neinhuis C. Superhydrophobic hierarchically structured surfaces in biology: Evolution, structural principles and biomimetic applications. *Phil. Trans. R. Soc. A* 2016, **374**, 20160191.
4. Bing W., Wang H., Tian L., Zhao J., Jin H., Du W., Ren L. Small structure, large effect: Functional surfaces inspired by Salvinia leaves. *Small* 2021, **2** (9), 2100079.
5. Amabili M., Giacomello A., Meloni S., Casciola C. M. Unraveling the Salvinia paradox: Design principles for submerged superhydrophobicity, *Adv. Mater. Interfaces* 2015, **2** (14), 1500248.
6. Bormashenko E., Musin A., Whyman G., Zinigrad M. Wetting transitions and depinning of the triple line. *Langmuir* 2012, **28**, 3460–3464.

7. Shanahan M. E. R., Sefiane K. Kinetics of Triple Line Motion during Evaporation, in *Contact Angle Wettability and Adhesion*, Mittal K. L., Ed., VSP, Leiden, The Netherlands; 2009; Vol. **6**, pp. 19–31.

8. Bormashenko E., Musin A., Zinigrad M. Evaporation of droplets on strongly and weakly pinning surfaces and dynamics of the triple line. *Colloids Surf., A* 2011, **385**, 235–240.

9. Bhushan B. Biomimetics. *Bioinspired Hierarchical-Structured Surfaces for Green Science and Technology, 3rd Ed*, Fabrication and Characterization of Micropatterned Structures Inspired by Salvinia molesta, Springer Series in Materials Science, Springer, Cham, Switzerland; 2019; Vol **279**.

10. Barthlott W., Mail M., Bhushan B., Koch K. Plant surfaces: Structures and functions for biomimetic innovations. *Nano-Micro Lett.* 2017, **9**, 23.

5 Shark Skin Effect

From the first glance, it seems that in order to manufacture the clean surfaces, we have to polish them in the best possible way and to create interfaces that are as smooth as possible. In a rather paradoxical way, nature chose a different way for preparing self-cleaning interfaces. The self-cleaning surface of lotus is rough on the micro- and nano-scales, as discussed in Chapter 2. What is necessary for the development of maximum velocity for self-propelled bodies fully immersed in water? Again, it seems that for the drag reduction, we have to polish the surface of bodies in the best possible way. And, again, nature chose a paradoxical alternative for the solution of this problem: the skin of the sharks, which are amazing swimmers, is rough and it is covered with the microscaled riblets. Although sharks are commonly described as cartilaginous fishes, they are in fact covered by numerous small dermal tooth-like elements termed placoid scales or denticles. There is a particular interest in the hydrodynamics of shark scales (squama); studies have shown that they can reduce skin friction drag and enhance hydrodynamic performance in separating flows. Theoretical, experimental and numerical studies have shown that idealized two-dimensional riblets can reduce drag by up to 10%. The shark skin effect is the reduction of the fluid drag during swimming at fast speeds and protection of its surface against biofouling. The presence of surface microstructure on the skin surface is responsible for this effect. The shark skin effect demonstrates enormous potential for shipbuilding and aircraft industries.

5.1 FLUID DRAG AND ITS ORIGIN

In fluid dynamics, drag (sometimes called fluid resistance) is a force acting opposite to the relative motion of any object moving with respect to a surrounding fluid.[1,2] Fluid drag comes in several forms, the most basic of which are pressure drag and friction drag (an excellent review of physics of fluid drag is found in References 1 and 2). Pressure drag is the drag associated with the energy required to move fluid out from in front of an object in the flow and then back in place behind the object. Much of the drag associated with movement through liquid is pressure drag, as the liquid in front of a body must be moved out and around the body before the body can move forward.[2]

$$F_{PD} = \frac{1}{2} C_D \rho v^2 A, \tag{5.1}$$

where ρ is the density of the liquid, v is the speed of the object relative to the fluid, A is the cross-sectional area and C_D is the drag coefficient, which is a dimensionless number (the values of C_D for different shapes of bodies are summarized in Reference 1).

DOI: 10.1201/9781003178477-5

The magnitude of pressure drag can be reduced by creating streamlined shapes. Friction or viscous drag, in turn, is caused by the interactions between the fluid and a surface parallel to the flow, as well as the attraction between molecules of the fluid (see the discussion in Section 1.15). Friction drag is similar to the motion of a sliding card deck across a table.[1,2] The frictional interactions between the table and the bottom card, as well as between each successive card, mimic the viscous interactions between molecules of fluid.[1,2] Moving away from the surface of an object in a fluid flow, each fluid layer has higher velocity until a layer is reached where the fluid has velocity equal to the mean flow. Fluids of higher viscosity – the attraction between molecules – have higher apparent friction between fluid layers, which increases the thickness of the fluid layer distorted by an object in a fluid flow. For this reason, more viscous fluids have relatively higher drag than less viscous fluids. A similar increase in drag occurs as fluid velocity increases. The drag on an object is, in fact, a measure of the energy required to transfer momentum between the fluid and the object to create a velocity gradient in the fluid layer between the object and undisturbed fluid away from the object's surface. For the simplest case of the sphere with radius R moving in the liquid, the viscous drag F_{VD} force is given by the Stokes Law:

$$F_{VD} = 6\pi R \eta v, \qquad (5.2)$$

where η is the viscosity of the liquid, discussed in detail in Sections 1.4 and 1.5. It is immediately recognized that both the pressure and viscous drug force opposing the motion grow with the velocity of the body (the drag force will play a central role in the theory of damped oscillations, to be addressed in Section 7.1.5). Hence, development of high velocities for the self-propelled bodies immersed in water is not a trivial task. At the same time, sharks are some of the fastest and most agile swimmers in the ocean. They are capable of reaching maximal speeds v_{max} of up to 60 miles per hour $\left(v_{max} \cong 100 \frac{km}{hour} \cong 28 \frac{m}{s}\right)$, which is faster than most other fish and marine mammals. How is it possible?

At least partially, the fast movement of sharks becomes possible due to the surface topography of the shark skin, reducing the viscous drag. The above discussion of friction drag, giving rise to Eq. (5.2), assumes that all neighboring fluid molecules move in the same relative direction and momentum transfer occurs between layers of fluid flowing at different velocities; in other words, Eq. (5.2) works for the laminar flow (laminar flow is characterized by fluid particles following smooth paths in layers, with each layer moving smoothly past the adjacent layers with little or no mixing). At low velocities, the fluid tends to flow without lateral mixing and adjacent layers slide past one another like playing cards, if a deck of cards analogy is again used.[1] Transition from laminar- to turbulent-flow regimes occurs near a Reynolds number around 2,900 for pipe flow (the Reynolds number $Re = \frac{\rho v D}{\eta}$ was introduced in Section 1.15). Let us estimate the Reynolds number for the flow occurring in the vicinity of the shark skin. Assuming for the typical size of the riblets forming the shark skin (see Figure 5.1) $D \cong 200 \mu m = 2 \times 10^{-4} m$, $\rho \cong 10^3 \frac{kg}{m^3}$, $\eta \cong 10^{-3} Pa \times s$ and $v \cong 20.0 \frac{m}{s}$, we estimate for the Reynolds number $Re \cong 4,000$. Thus, the conditions for the turbulent flow in the vicinity of the swimming shark are created and vortices

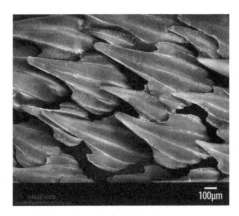

FIGURE 5.1   Typical structure of the shark skin riblets. The scale bar is 100 μm.

typical for the turbulent flow are created in the area adjacent to the shark skin. It was suggested that the small riblets that cover the skin of fast-swimming sharks, depicted in Figure 5.1, work by impeding the cross-stream translation of the stream-wise vortices in the viscous sublayer.[2] The mechanism by which the riblets/denticles interact with and impede vortex translation is complex and the entirety of the phenomena is not yet fully understood. Impeding the translation of vortices reduces the occurrence of vortex ejection into the outer boundary layers as well as the momentum transfer caused by tangling and twisting of vortices in the outer boundary layers.[2] Riblets increase the surface area.

In the turbulent-flow regime, fluid drag typically increases dramatically with an increase in surface area owing to the shear stresses at the surface acting across the new, larger surface area. However, in a very paradoxical way, as vortices form above a riblet surface, they remain above the riblets, interacting with the tips only and rarely causing any high-velocity flow in the valleys of the riblets. Since the higher velocity vortices interact only with a small surface area at the riblet tips, only this localized area experiences high-shear stresses.[2] The low-velocity fluid flow in the valleys of the riblets produces very low-shear stresses across the majority of the surface of the riblet, thus resulting in the reducing of the drag force.[2]

It should be mentioned that the shark skin riblets/denticles can be round and peg-like, diamond-shaped with ridges and pointed posterior margins, or they can lack surface ornamentation and have a smooth and rounded posterior edge.[3] The geometry of the shark skin pattern is complicated and the analysis of the flows emerging from the contact with the shark skin is possible only with the sophisticated numerical methods.[4]

There is substantial disagreement between comparable studies of the flow over three-dimensional shark scales; some studies report poor hydrodynamic performance of denticles compared, some report similar performance to idealized riblets and some report better performance than longitudinal riblets, where skin friction drag is reduced by up to 30%.[4] The broad range of contradictory results presented in the literature are in part due to differences in denticle geometries and spacings, arising from differences in denticle samples and manufacturing techniques.[4] Thus, we necessarily conclude that a comprehensive study on the effects of denticle geometry and spacing on skin friction has not been carried out.[4]

It was demonstrated that shark skin topography improves not only hydrodynamic but also aerodynamic characteristics of the surfaces.[5] Denticles not only decrease the frag force but also generate lift, resulting in high lift-to-drag ratio improvements.[5] In spite of the fact that physical mechanisms of the "shark-skin" effect are not completely understood, the 3D printing technologies enabling manufacturing of the shark-skin-like surfaces were already reported. Using 3D printing, thousands of rigid synthetic shark denticles were placed on flexible membranes in a controlled, linear-arrayed pattern.[6] Compared with a smooth control model without denticles, the 3D printed shark skin showed increased swimming speed with reduced energy consumption under certain motion conditions.[6]

Bullets

- The shark skin effect is the reduction of the fluid drag during swimming at fast speeds and protection of its surface against biofouling.
- The presence of surface microstructure on the skin surface is responsible for this effect.
- The shark skin effect demonstrates enormous potential for shipbuilding and aircraft industries.

EXERCISES

1. The Sanbar Shark (*Carcharhinus plumbeus*) moves relatively to water with a velocity of $v = 10.0 \frac{m}{s}$. We assume that the cross-section of the shark is the circle with the radius $R = 25$ cm. Assume that the drag coefficient $C_D \cong 0.4$ (for the values of the drag coefficient, see Reference 1). The density of seawater is $= 1.03 \times 10^3 \frac{kg}{m^3}$, the viscosity of sea water at room temperature $\eta \cong 9.0 \times 10^{-4} Pa \times s$.

    a. What the Reynolds number, related to the motion of the shark as an entire body? What kind of water motion is expected in the vicinity of the shark?

    The Reynolds number is given by $Re = \frac{\rho v R}{\eta}$; substitution of the numerical values of the parameters yields $Re \cong 2.7 \times 10^6$. This means that the turbulent motion of water will be inspired by the motion of the shark.

    b. What is the value of the pressure drag acting on the shark?

    The pressure drag is given by $F_{PD} = \frac{1}{2} C_D \rho v^2 A \cong \frac{\pi}{2} C_D \rho v^2 R^2$. Actually, this formula supplies not more than a very rough estimation of the pressure drag, due to the complicated hydrodynamic events accompanying the motion of the shark (see Reference 1). Anyway, let us exploit the aforementioned expression for a sake of a very rough estimation of the pressure drag. Substitution of the numerical values of the parameters yields $F_{PD} \cong 3.75 \times 10^3$ N.

2. Solid sphere with the radius $R = 25$ cm moves in a sea water at a room temperature with a velocity of $v = 10.0 \frac{m}{s}$. Estimate the viscous drag force acting on the sphere; the viscosity of sea water at room temperature $\eta \cong 9.0 \times 10^{-4} Pa \times s$. Compare the calculated viscous drag force with the pressure drag acting on the shark, estimated in the previous problem.

REFERENCES

1. Pritchard P. J., Leylegian L. C. Fox and McDonald, in *Introduction to Fluid Mechanics*, *8th ed.*, John Wiley and Sons, Inc., Hoboken, NJ, USA; 2011; Chapter 9.7, pp. 445–460.
2. Dean B., Bhushan B. Shark-skin surfaces for fluid-drag reduction in turbulent flow: A review. *Phil. Trans. R. Soc. A* 2010, **368**, 5737.
3. Domel A. G., Domel G., Weaver J. C., Saadat M., Bertoldi K., Lauder G. V. Hydrodynamic properties of biomimetic shark skin: Effect of denticle size and swimming speed. *Bioinspir. Biomim.* 2018, **13**, 056014.
4. Lloyd C. J., Mittal K., Dutta S., Dorrell R. M., Peakall J., Keevil G., Burns M. D. Multi-fidelity modelling of shark skin denticle flows: Insights into drag generation mechanisms. *R. Soc. Open Sci.* 2023, **10**, 220684.
5. Domel A. G., Saadat M., Weaver J., Haj-Hariri H., Bertoldi K., Lauder G. V. Shark denticle-inspired designs for improved aerodynamics. *J. R. Soc. Interface* 2018, **15**, 20170828.
6. Wen L., Weaver J. C., Lauder G. V. Biomimetic shark skin: Design, fabrication and hydrodynamic function. *J Exp. Biol.* 2014, **217** (10), 1656–1666.

6 Gecko Effect and Novel Adhesives

Geckos are carnivorous lizards that have a wide distribution, found on every continent except Antarctica. Geckos are found in warm climates throughout the world. They range from 1.6 to 60 cm. Geckos have the largest mass and have developed the most complex hairy attachment structures among climbing animals that are capable of smart adhesion – the ability to cling to different smooth and rough surfaces and detach at will. Geckos can run rapidly on walls and ceilings, requiring high friction forces (on walls) and adhesion forces (on ceilings), with typical step intervals of ca. 20 ms. Obviously, they demonstrate a kind of "non-glue" sticking to solid surfaces. We address in this chapter the following questions: (i) what is the physics of this dry adhesion? (ii) how thus unusual adhesion may be reproduced artificially? It is already agreed that Geckos make use of about three million microscale hairs (setae) (about 14,000/mm^2) that branch off into nanoscale spatulae, about three billion spatulae located on their feet for smart attachment/detachment to/from solids. The so-called division of contacts provides high dry adhesion. Multiple-level hierarchically structured surface construction provides the Gecko with the compliance and adaptability to create a large real area of contact with a variety of surfaces that, in turn, yields the large values of adhesion forces. Again, the hierarchical structure of these hairs makes the effect possible.

6.1 GECKO EFFECT: MACROSCOPIC APPROACH

Although there are over 1,000 species of Geckos that have attachment pads of varying morphology, the Tokay Gecko (Gekko Gecko), which is native to southeast Asia, has been the main focus of scientific research.[1,2] Two variants of Tokay Geckos exist: a red-spotted and a black-spotted. In the red-spotted Tokay Gecko (depicted in Figure 6.1), these spots range from light yellow to red and overlay a bluish or grayish

FIGURE 6.1 Red-spotted Tokay Gecko is shown.

DOI: 10.1201/9781003178477-6

body. Their coloration is important for camouflage and they can actually lighten or darken their skin color to better blend in with their environment. The Tokay Gecko can cast off its tail in defense, breaking it off at several sections.

The cast-off portion of the tail continues to move violently for several minutes, giving the Gecko time to escape. The Gecko's tail regenerates in approximately three weeks, although it typically does not grow as long as the original. There have been concerns regarding locally decreasing populations due to overexploitation of Tokay Gecko for traditional Chinese medicine.[2]

The Tokay Gecko is the second largest Gecko species, attaining lengths of approximately 0.3–0.4 m and 0.2–0.3 m for males and females, respectively. They have a distinctive blue or gray body with orange or red spots and can weigh up to 0.3 kg. These have been the most widely investigated species of Gecko due to the availability and size of these creatures.[1,2] Hierarchical structure of Tokay Gecko feet is noteworthy: one body with four feet, each foot with five toes, each toe with $\cong 20$ rows of sticky lamellae, each lamella with many setal arrays consisting of thousands of setae, which amounts to $\cong 200,000$ setae per toe, and each seta consisting of hundreds to 1,000 spatulae at its end, as shown in Figure 6.2.

Let us try to understand the mechanism of attachment/detachment of Gecko to/from solid surfaces. We start from the macroscopic approach to adhesion. The classical macroscopic approach to adhesion is supplied by the Johnson–Kendall–Roberts (JKR) theory, which was developed as an extension of the famous Hertz theory of the contact of two elastic bodies.[3,4] Consider two elastic spheres of radius R_1 and

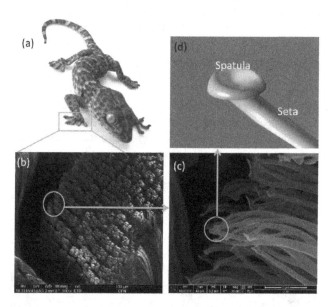

FIGURE 6.2 Hierarchical structure of the Gecko foot is depicted. (a) Optical image of a Tokay Gecko. (b) SEM image of arrays of setae on the Gecko's toes. (c) SEM image of spatulae and setae in (b). (d) Schematic of the Gecko's setae and spatula, where the spatulae are shaped like nanometer-sized suction cups.

R_2 pressed together under a load P_0; the radius a of the circle of contact is given by Eq. (6.1):

$$a^2 = \frac{3\pi}{4}(k_1 + k_2)\frac{R_1 R_2}{R_1 + R_2}P_0, \qquad (6.1)$$

where $k_i = \frac{1-v_i^2}{\pi E_i}$; $i = 1, 2$, v_i is the Poisson ratio and E_i the Young modulus of each material. Consider now an extremely important concept for materials science: *the Young modulus and the Poisson ratio exhaust the elastic properties of the elastic material*; in other words, any characteristic of the isotropic elastic body should be necessarily expressed via the elastic modulus E and the Poisson ration v. Eq. (6.1) illustrates this idea: indeed, the contact radius a is expressed via the Poisson ratio v_i and the Young modulus E_i of the contacting media. Resulting from local compression near to the contact region, distant points in the two spheres approach each other by a distance δ given by

$$\delta^3 = \frac{9}{16}\pi^2(k_1 + k_2)^2\frac{R_1 + R_2}{R_1 R_2}P_0^2. \qquad (6.2)$$

Experiments carried out with the solids pressed one to another demonstrated that attractive surface forces are acting between pressed solids. These contact forces are of little significance at high loads; however, they became increasingly important as the load was reduced toward zero. Surface attraction may be interpreted in terms of surface energy.

Consider two elastic spheres in contact under zero external load. Attractive forces between the surfaces produce a finite contact radius, denoted a, a balance eventually being established between stored elastic energy and lost surface energy. The loss in surface energy U_s is given by

$$U_s = -\pi a^2 \hat{\gamma}, \qquad (6.3)$$

where $\hat{\gamma}$ is an effective surface energy per unit area. For two dissimilar surfaces, the effective surface energy is given by

$$\hat{\gamma} = \gamma_1 + \gamma_2 - \gamma_{12}, \qquad (6.4)$$

where γ_1 and γ_2 are the surface energies of each surface, and γ_{12} is the interface energy. When the two surface materials are identical, $\gamma_{12} = 0$, the interface energy is zero and the effective surface energy reduces to $\hat{\gamma} = 2\gamma$. The force associated with the surface energy U_s is given by Eq. (6.5).

$$F_s = -\frac{dU_s}{dx}. \qquad (6.5)$$

where x is the movement of the bodies and is approximately the same as δ which is given by Eq. (6.2) and it cannot be worked out exactly because the attractive surface forces disturb the stress distributions in the bodies. For the sake of a very rough approximation, it may be assumed

$$x \approx a^2 \frac{R_1 + R_2}{R_1 R_2}. \tag{6.6}$$

Eq. (6.6) leads to extremely important scaling conclusion, namely: $x \sim \frac{a^2}{R}$, which will be exploited below for explanation of the Gecko adhesion. Combining Eqs. (6.4)–(6.6) yields:

$$F_s = \pi \hat{\gamma} \frac{R_1 R_2}{R_1 + R_2}. \tag{6.7}$$

This force acts in addition to the ordinary load P_0 between surfaces and the simple analysis shows that it may be related to the geometry and energy of the contacting surfaces. It immediately results from Eq. (6.7) that force of adhesion F_s between convex surfaces does not depend upon the elastic moduli of the materials. This is quite a surprising result. The elastic modulus influences the contact radius a, but, however, it can be seen from Eqs. (6.3) and (6.6) that both the surface energy and the elastic work vary as a^2, so that the force of adhesion is independent of the contact radius a and hence of the elastic modulus. The more rigorous analysis provided the magnitude of the adhesive force but it did not change this conclusion. An accurate mathematical treatment of the problem yields for the modulus of the adhesion force:

$$F_s = \frac{3}{2} \pi \hat{\gamma} \frac{R_1 R_2}{R_1 + R_2}. \tag{6.8}$$

Again, the adhesion force is independent of the elastic modulus. For identical spheres in contact ($R_1 = R_2 = R$), we obtain:

$$F_s = \frac{3}{4} \pi \hat{\gamma} R. \tag{6.9}$$

Eq. (6.9) leads to a paradoxical conclusion[5–7]: the pull-off force for a large contact is smaller than the sum of the pull-off forces for many smaller contacts together, covering the same apparent contact area. This concept is known as the "principle of contact splitting". This principle is explained as follows: the number of equally sized spherical contacts n is proportional to the contact area $S_{contact}$, which itself is proportional to the square of a length. Considering Eq. (6.9) and the scaling relation $x \sim \frac{a^2}{R}$ which leads to an important conclusion, it follows that for the fixed x, the pull-off force F_s is proportional to the square root of the number of contacts, according to the following scaling reasoning:

$$F_s \sim R \sim \sqrt{S_{contact}} \sim \sqrt{n}. \tag{6.10}$$

The aforementioned analysis is applicable for the analysis of the Gecko adhesion, when the Gecko spatula are simplified as cylinders, each terminated with a hemispherical end of radius R.

6.2 PHYSICAL ORIGIN OF THE GECKO EFFECT: VAN-DER-WAALS FORCES AND THEIR ROLE IN THE DRY ADHESION

6.2.1 PHYSICAL AND CHEMICAL INTERACTIONS IN THE MATERIALS SCIENCE—SHORT-RANGE AND LONG-RANGE INTERACTIONS

We already mentioned the importance of hierarchical structure of Gecko feet, crucial for its adhesion to solid surfaces. What is known about the adhesion of Gecko feet to different surfaces on various levels? On the smallest level of spatula, the adhesion force is established as ca. 10 nN; the maximum friction force of a single seta (with 100–1,000 spatulae) is $\cong$ 200 μN, and the adhesive force is 20–40 μN. The researchers agreed that the friction forces are much higher than the adhesion forces.[8] It is also agreed that both adhesion and friction forces emerge from the van der Waals intermolecular forces. Let us acquaint briefly these forces, which are so important in the modern materials science. We start from the general classification of interactions in materials science. Scientists distinguish between physical and chemical interactions.[9] Chemical interatomic forces form chemical bonds within a molecule. Excellent introduction to this classification is found in References 9 and 10. Chemical interatomic forces between atoms result in an atomic aggregate with sufficient stability to form a chemical bond within a molecule.[9,10] Three main types of chemical bonds are distinguished: they are metallic, covalent and electrostatic (ionic) bonds. These bonds are responsible for the formation of molecules. Molecules, in turn, interact with one another. Now we came to the second general concept, which is of primary importance for the materials science: *all the properties of bulk material can be determined by the number and types of molecules it contains and their arrangement in space with respect to each other, which, in turn, defines interactions between the molecules.*[9,10] These interactions arise from physical forces, which are different from chemical interactions/forces which give rise to chemical bonds, which form a molecule.[9,10] Chemical covalent interactions change completely electron charge distributions of interacting atoms and merge.[9,10] Physical binding only disturbs the pristine charge distributions of interacting molecules. Physical interactions lack the directionality and stoichiometry of chemical covalent bonds.[9,10]

Physical interactions can be as strong as covalent ones to hold the molecules together in solids and liquids at room temperatures. Accurate analysis of physical interactions necessarily should consider the quantum mechanics aspects of these intermolecular interactions; however, this is a still very challenging computational task. Thus, it is convenient to separate the total intermolecular energy in a number of parts, namely: a Coulomb, polar, induced-polarized, van der Waals, hydrogen-bonding, hydrophobic and hydrophilic interactions.[9,10] It should be emphasized that all physical forces acting between molecules are essentially electrostatic (gravitational interactions may be neglected). Intermolecular interactions may be very generally represented as follows:

$$U(r) = -C \frac{x_1 x_2}{r^n}, \tag{6.11}$$

where C is an interaction constant, x is the amount of the molecular property involved in the interaction, r is the center-to-center separation distance between the molecules and n is the integer. Now we calculate the total energy of the given molecule with diameter d_m with other molecules confined within the system of the radius L. We denote this energy U_{int}^{total} (see **Appendix 1.A**):

$$U_{int}^{total} = \int_{d_m}^{L} U(r)4\pi r^2 dr = -\frac{4\pi C x_1 x_2 \rho}{(n-3)d_m^{n-3}}\left[1-\left(\frac{d_m}{L}\right)^{n-3}\right]. \tag{6.12}$$

For the details of the calculation of Eq. (6.12), see References 9 and 10 and **Appendix 1.A**. Eq. (6.12) is of primary conceptual importance for the materials science. It is reasonable to suggest that the dimension of the system L is much larger than the typical diameter of the molecule d; in other words, $L \gg d$ takes place. Thus, the total energy of interaction of the fixed molecule with its neighbors will be independent on the dimension of the system/box only when $n > 3$ is assumed (see Eq. (6.12)). Recall the already introduced concept: *all the properties of bulk material can be determined by the number and types of molecules it contains and their arrangement in space with respect to each other, which, in turn, defines inter-actions between the molecules.* Thus, the total energy of interaction of the given molecule is expected to define the bulk properties of the medium. Thus, the bulk properties of the gases, liquids and solids do not depend on their volumes only for values of n greater than 3. In this case, the intermolecular forces are not expected to extend over large distances but depend only on the forces between molecules in close proximity to each other. These forces are labeled the *"short-range forces"*. Usually, intermolecular force potentials do indeed predict that n usually exceeds 3 and it is for this reason that the intrinsic bulk properties of solids, liquids and gases do not depend on the volume of the material or on the size of the container (unless they are extremely small) but only on the forces between molecules in close proximity to one another. The forces for which $n < 3$ takes place are called *the long-range forces*. The well-known fundamental long-range forces are gravitational and Coulomb ($= 1$ in Eq. (6.11)) and the forces between magnetic or electric dipoles ($n = 3$ in Eq. (6.11)). The situation when $n = 3$ is of a special interest for the materials science. The situation when n exactly equals 3 corresponds to the force acting between magnetic or electric dipoles. The total energy of their interaction continues to increase with r as $\ln r$. This increase is very weak due to the fact that the logarithm is the weak function. However, it increases and it is for this reason that magnetic dipoles can mutually align themselves along the same direction in a magnetic material and why we can feel magnetic forces (such as earth's magnetic field) over very large distances.[10] Such interactions also give rise to a long-range ordering of molecules possessing the electrical dipole moment.[10]

Very strong long-range forces are the electrostatic Coulomb forces and they are responsible for the formation of the ionic crystals. For bi-ionic interactions of two ions $q_1 = z_1 e$, $q_2 = z_2 e$ (where z_1 and z_2 are ionic valences of ions, $e = 1.602 \times 10^{-19}$ C is the elementary charge), the potential of interaction is given by Eq. (6.13):

$$U(r) = -\frac{z_1 z_2 e^2}{4\pi\varepsilon_0\varepsilon_r r}, \tag{6.13}$$

where $\varepsilon_0 = 8.85 \times 10^{-12} \frac{F}{m}$ is the vacuum permittivity and ε_r is the relative permittivity or dielectric constant of the medium (ε_r is the ratio of the electric strength in a vacuum E_0 to the electric field strength with a medium $\varepsilon_r = \frac{E_0}{E}$). Coulomb electrostatic forces govern interaction in the ionic crystals, such as NaCl, KBr and NaF. And these forces are definitely "long-range" ones ($n = 1$); thus, the bulk properties of ionic crystals are expected to depend on the volume of the crystal or on the size of the container, confining the crystal. However, it is well-known that the bulk properties of ionic crystals are independent on their volume; in other words, they are intensive and not extensive. How is it possible? In order to explain intensive properties of the ionic crystals, we have to consider the interaction of the fixed ions with *all of its neighbors* and not only with the nearest ones (and this is quite reasonable, due to the long-range nature of the Coulomb interactions). If we examine the NaCl crystal lattice carefully, each Na^+ ion has six nearest Cl^- ions at $r = 0.276$ nm, 12 next-nearest neighbor Na^+ ions at $r_2 = \sqrt{2}r$, eight more Cl^- ions at $r_2 = \sqrt{3}r$ and so on.[9,10] The total interaction energy of the given ion in the ionic crystal lattice is calculated by summing the pair potentials as follows:

$$U_{tot} = \frac{e^2}{4\pi\varepsilon_0 r}\left(6 - \frac{12}{\sqrt{2}} + \frac{8}{\sqrt{3}} - \ldots\right) = -1.748\frac{e^2}{4\pi\varepsilon_0 r} = -1.46 \times 10^{-18}\,\text{J}. \quad (6.14)$$

The constant $M = 1.748$ for an ionic NaCl crystal is known as the Madelung constant and varies numerically for different ionic crystals. We recognize infinite mathematical series in Eq. (6.14). The fundamental mathematical questions of convergence and uniqueness of the sum of these series are deeply discussed in Reference 11. The existence of intensive properties of ionic crystals, such as their density, supplies the *physical proof* of the convergence of the series, appearing in Eq. (6.14). Is the energy calculated with Eq. (6.14) large or small? In order to answer this question, we have to define the reasonable scale. *Now we come to one concept, which is extremely important for physics and materials science: the statement "the energy is large", or "the energy is small" is senseless until the appropriate scale is defined. The notions of "large" and "small" are not absolute but relative.* What is the reasonable energy scale in this case (consider that energy may be compared only to energy)? Eq. (6.14) predicts the potential energy of ions in the ionic crystals; it is plausible to compare the potential energy of the ion to its kinetic energy, which is, in turn, quantified with the temperature of the crystal denoted as T. According to the equipartition theorem serving as a basis for the classical thermodynamics and statistical physics, in thermal equilibrium, energy is shared equally among all of its various forms.[12] Very roughly, the thermal energy which falls on the degree of freedom of the particle/molecule is estimated as $k_B T$, where k_B is the Boltzmann constant and T is the temperature of the system in thermal equilibrium.

Calculation of the total energy of the ion in the NaCl lattice with Eq. (6.14) yields the estimation $U_{tot} \cong 350\, k_B T$. This means that the potential energy of the ion in the NaCl lattice is much larger than its kinetic energy; actually, this interrelation between the potential and kinetic energies explains to a large extent the behavior of solids in which atoms/molecules are stably located in their equilibrium positions and do not like to be removed far from their equilibrium locations. The value of the Madelung constant varies between 1.638 and 1.763 for crystals composed of monovalent ions such as NaCl and CsCl, rising to about 5 for monovalent-divalent ion pairs such as CaF_2, and higher for multivalent ions such as SiO_2 and TiO_2.[10,11] The interaction energy is negative and it is of the same order for all isolated pairs.

6.2.2 INTERACTIONS BETWEEN POLAR AND NON-POLAR MOLECULES

Most of molecules carry no net charge. The electrical charge excess results in a giant increase of the energy of molecule and it is thermodynamically unfavorable. So, how do molecules interact? Many of molecules possess an electric dipole. This takes place, when positive and negative charges are separated within a molecule as it occurs for the water molecule H_2O, depicted in Figure 6.3A. The electric dipole moment has a magnitude $\tilde{p} = ql$ (where q is a point charge and l is the separation between the charges), as shown in Figure 6.3B.

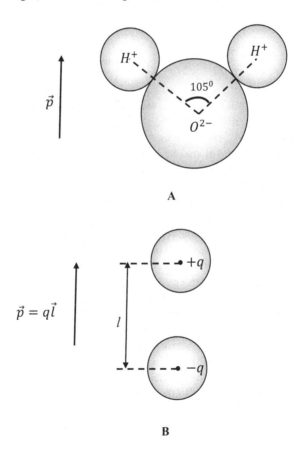

FIGURE 6.3 (**A**) Dipole molecule of the water molecule is illustrated. (**B**) Definition of the dipole moment $\vec{p}$. l is the distance between the centers of mass of the electrical charges q and $-q$.

TABLE 6.1

Dipole Moments (in Debye) and Polarizability of Molecules α

Molecule	Dipole Moment, in Debye, 1D $1D = 3.336 \times 10^{-30}$ Cm	Polarizability, α, $(4\pi\varepsilon_0)\text{Å}^3 =$ $(4\pi\varepsilon_0) \times 10^{-30} \text{m}^3$
H_2O	1.85	1.45–1.48
$NaCl$	8.5	
$CsCl$	10.4	
CCl_4	0	10.5

The electric dipole moment $\vec{p}$ is a vector and it is directed from the negative charge to the positive one, as shown in Figure 6.3B. For example for two electrical charges $q = \pm e$, separated by $l = 0.1$ nm, the dipole moment $\tilde{p}$ is $\tilde{p} = 1.602 \times 10^{-19}$ C$\times 10^{-10}$ m $\cong 1.6 \times 10^{-29}$ Cm $= 4.8$D. The unit of dipole moment is Debye, where $1\text{Debye} = 1\text{D} = 3.336 \times 10^{-30}$ Cm. Dipole moments of some of the molecules are supplied in Table 6.1. It is important that molecules of some of compounds (CCl_4) gas zero dipole moment.

Now consider two polar molecules mutually oriented as shown in Figure 6.4. Figure 6.4 illustrates the mutual orientation of two dipoles $\vec{p}_1, \vec{p}_2$ (dipolar molecules), separated by distance r (it is latently supposed that this distance is larger than the characteristic size of the addressed pair of dipoles $\vec{p}_1, \vec{p}_2$; the moduli of the dipoles are denoted $\tilde{p}_1$ and $\tilde{p}_2$ correspondingly). The triad of angles $\theta_1, \theta_2, \varphi$, shown in Figure 6.4 define the mutual orientation of the dipoles.[9,10] The potential energy of electrostatic interaction between the dipoles is given by Eq. (6.15):

$$U_{el}(r) = -k \frac{\tilde{p}_1 \tilde{p}_2}{r^3} (2\cos\theta_1\cos\theta_2 - \sin\theta_1\sin\theta_2\cos\varphi), \qquad (6.15)$$

where ε is the dielectric constant of the medium in which the dipoles $\vec{p}_1, \vec{p}_2$ are embedded, $k = \frac{1}{4\pi\varepsilon_0}$.

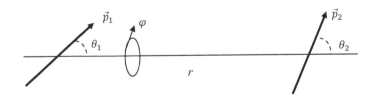

FIGURE 6.4   Parameters of the dipole-dipole electrostatic interaction; the dipole separation r is much larger that the characteristic size of the molecule; θ_1, θ_2 are the polar angles of the dipoles $\vec{p}_1, \vec{p}_2$ and φ is the rotation angle of the dipole rotation; the triad of angles $\theta_1, \theta_2, \varphi$ define the mutual orientation of the dipoles.

Eq. (6.15) works only for fixed dipole orientations. However, actually the polar molecules are mobile and may rotate, so that the "angle-averaged potential" should be applied for quantification of the dipole-dipole interactions. Keesom performed statistical calculations averaging over all mutual orientations of dipoles weighted by the Boltzmann factor and derived the average potential energy of interaction between a pair of dipoles $\vec{p}_1$, $\vec{p}_2$ embedded into a medium with a dielectric constant ε:

$$U_K(r) = -\frac{\tilde{p}_1^2 \tilde{p}_2^2}{3(4\pi\varepsilon_0\varepsilon)^2 k_B T r^6}. \tag{6.16}$$

Eq. (6.16) works when $k_B T > \frac{\tilde{p}_1 \tilde{p}_2}{4\pi\varepsilon_0\varepsilon r^3}$ takes place and it is called *the Keesom interaction*. Actually, the Keesom interactions are the part of the van der Waals interactions to be addressed below.[9,10]

As it is seen from Table 6.1, some of molecules, such as CCl_4, have no permanent dipole moment (in other words, their permanent dipole moment is zero). It is expected that electrostatic interaction between these molecules is impossible; however, this conclusion is erroneous, due to the fact that all molecules are polarizable. When a non-polar molecule is subjected to an electric field, the electrons in the molecule are displaced from their ordinary positions so that electron clouds and nuclei are attracted in opposite directions. The charges inside the molecule shift, a dipole is induced, and the molecule has a temporarily induced dipole moment $\tilde{p}_{ind}$. This moment is proportional to the electric field E experienced by the molecule:

$$\tilde{p}_{ind} = \alpha E, \tag{6.17}$$

where α is the molecular polarizability (see also Exercise 6.11). For a non-polar molecule in free space, the polarizability α is supplied by Eq. (6.18);

$$\alpha = 4\pi\varepsilon_0 r^3, \tag{6.18}$$

where r is the radius of the molecule. It is immediately seen from Eq. (6.18) that *there is a direct proportionality between the polarizability and the volume of the molecule.*[9] In CGS units, the dimension of polarizability is $[\alpha] = cm^3$ and it is in the range of $10^{-24} cm^3 = 1 (A^0)^3$. And, indeed, Eq. (6.18) may be very roughly used for estimation of the polarizability of molecules (see Table 6.1 for polarizability of some molecules).[9]

When we speak of the polar molecules, their polarizability is built from two components, one of which is given by Eq. (6.18) and the second is the so-called orientation polarizability. How does it emerge? A free rotating dipolar molecule has a time-averaged zero dipole moment. However, when an electric field E is applied, the Boltzmann-averaged orientations of the rotating dipoles change the average dipole moment so that an orientation polarizability α_{orient} of the molecule is formed. It is given by Eq. (6.19):

$$\alpha_{orient} = \frac{\tilde{p}^2}{3k_B T}. \tag{6.19}$$

The total polarizability of polar molecules is given by combining Eq. (6.18) and Eq. (6.19):

$$\alpha_{total} = 4\pi\varepsilon_0 r^3 + \frac{\tilde{p}^2}{3k_B T}. \tag{6.20}$$

General expression describing interaction between dipolar molecules was obtained by Debye. Consider two molecules possessing dipole moments $\tilde{p}_1$ and $\tilde{p}_2$ and total polarizabilities α_1 and α_2. Interaction between these molecules embedded into the medium with dielectric constant ε is given by the Debye equation:

$$U_D(r) = -\frac{\tilde{p}_1^2 \alpha_2 + \tilde{p}_2^2 \alpha_1}{(4\pi\varepsilon_0\varepsilon)^2 r^6}. \tag{6.21}$$

The interactions described by Eq. (6.21) are often called Debye-induced dipole interactions. It is noteworthy that the Keesom interactions, described by Eq. (6.16), may be derived from Eq. (6.21) by substitution $\alpha_i = \frac{\tilde{p}_i^2}{3k_B T}$, $i = 1, 2$. Both Keesom and Debye interactions decrease with the inverse six power of the separation distance. And both of them contribute to the van der Waals interactions to be addressed in the next section.

6.2.3 Van Der Waals Interactions

The concept of intermolecular interactions was formulated by van der Waals, who studied deviations of behavior of real gases from that predicted by the equation of ideal gas:

$$PV = nRT, \tag{6.22}$$

where P is pressure, V is volume $R = 8.3145\,\mathrm{J\,K^{-1}mol^{-1}}$ is the gas constant (the gas constant is defined as $R = N_A k_B$, where $N_A = 6.02214076 \times 10^{23}\,\mathrm{mol^{-1}}$ is the Avogadro constant) and T is the absolute temperature). It was experimentally established that real gases deviate from behavior, predicted by Eq. (6.22). The most obvious experimental deviation is supplied by the phase changes; indeed, according to Eq. (6.22), no phase changes are expected for ideal gases for an entire range of pressures and temperatures, and it is well-known that real gases demonstrate phase transitions, when pressed or cooled. In order to explain this discrepancy, van der Waals in 1873 suggested the equation, which predicts a behavior of real gases in a much more realistic way:

$$\left(P + \frac{an^2}{V^2}\right)(V - nb) = nRT, \tag{6.23}$$

where a is constant which accounts for an attraction between molecules $\left([a] = \frac{\mathrm{Pa \times m^6}}{\mathrm{mol^2}}\right)$, and b, in turn, is a constant emerging from the repulsion between molecules $\left([b] = \frac{\mathrm{m^3}}{\mathrm{mol}}\right)$.

The second contribution to the ideal gas equation, i.e., nb is easy to understand: a gas can't be compressed all the way down to zero volume, so we've limited the volume a minimum value of nb at which the pressure goes to infinity. The constant b then represents the minimum volume occupied by a molecule, when it is "touching" all his neighbors. For the simple explanation of the physical meaning of the constant a, see Reference 13. As a matter of fact, the constant a is responsible for the term $\frac{an^2}{v^2}$ appearing in Eq. (6.23), which accounts for the *attractive intermolecular forces, known as van der Waals forces*, which cause an attraction between the gas molecules and restrict them to hit the walls of the container, thus decreasing the measured gas pressure from that predicted by the ideal gas (Eq. (6.22)). Actually, van der Waals Eq. (6.23) already contains the information about the phase transformations occurring into gases under their pressing or cooling.

Now we adopt the microscopic approach, and we'll try to understand how the macroscopic constants a and b appearing in Eq. (6.21) arise from microscopic considerations. Attractive intermolecular forces (known also as van der Waals forces) are built from three kinds of interactions, namely the Keesom forces between permanent diploes (see Eq. (6.16)), the Debye forces acting between induced dipoles (see Eq. (6.21)) and London (dispersion) interactions. The Keesom and Debye forces may be fully interpreted in terms of classical electrostatics. Contrastingly, the London dispersion forces are different and quantum mechanical in their nature. London demonstrated that non-polar molecules are non-polar only when viewed over a sufficiently long period of time, and their time-averaged distribution of electrons is symmetrical; if an instantaneous photograph of such a molecule is made, it would show, that at a given instant moment, the oscillation of the electrons around the nucleus, results in distortion of the electron cloud sufficient to cause a temporary dipole moment. The electron circulates with extremely high frequency ($f \sim 10^{16}\,\mathrm{Hz}$) and at every instant the molecule is therefore polar, but the direction of this polarity changes with a high frequency (the dipole moment is a vector (!), as shown in Figure 6.3). The dipole moment of the molecule is rapidly changing its direction and therefore averages out to zero over a short period of time; however, these quickly "rotating" dipoles produce an electric field which then induces dipoles in neighboring molecules.[9,10] The result of this induction is usually an attractive force acting between induced non-polar or dipolar molecules.[9,10] If the interacting molecules are of the same kind, the potential of the dispersion (London) interaction between the molecules $U_L(r)$ is given by Eq. (6.24).

$$U_L(r) = -\frac{3}{4}\frac{\alpha^2 I}{(4\pi\varepsilon_0)^2 r^6}, \tag{6.24}$$

where I is the first ionization potential of the molecule. Quite remarkably, the Keesom interactions (Eq. (6.16)), the Debye interactions (Eq. (6.21)) and the London interactions (Eq. (6.24)) decrease with the distance as $U(r) \sim \frac{1}{r^6}$. The London (dispersion) forces are, perhaps, most important contribution to the total van der Waals interactions, because they are always present (in contrast to other types of interactions, that may or may not be present, depending on polarization properties of the molecules).[9,10] Now we came to one extremely *important concept of the modern materials science:*

the London forces play a role in adhesion (including the Gecko adhesion), sur-face tension, wetting, etc. We already mentioned in Sections 1.3–1.5 that the values of the surface tension of the majority of organic liquids are close and this is due to the dominating nature of the London forces in the total van der Waals interactions.[14] The physical properties of the London forces are summarized in Reference 10:

1. The London/dispersion forces are long-range forces and can be effective from large distances (greater than 10 nm) down to interatomic spacing (about 0.2 nm).
2. The London forces may be repulsive or attractive.
3. Dispersion forces not only bring molecules together but also mutually orient them, though the orienting effect is weaker that with dipolar interactions.
4. Dispersion forces are not additive; in other words, the force between two bodies is affected by the presence of other bodies nearby.

Let us estimate the strength of the London (dispersion) forces. Assuming $\frac{\alpha}{4\pi\varepsilon_0} = 1.5 \times 10^{-30} \, \text{m}^3$, $I \cong 10^{-18} \, \text{J}$ and $r \cong 0.3 \, \text{nm}$ we estimate from Eq. (6.24) $|U_L(r)| \cong 4.6 \times 10^{-21} \, \text{J} \cong 1.0 \, k_B T$. This is a very respectable energy. Thus, while small non-polar atoms and molecules such as argon and methane are gaseous at room tem-perature and pressure, larger molecules such as high molecular weight hydrocarbons are liquids and solids held together solely by London (dispersion) forces.[10]

And it turns out that the London forces are responsible for the unusual adhesion of Gecko, enabling it hanging on and motion along vertical surfaces. The rapid switch-ing between Gecko foot attachment and detachment was analyzed in Reference 8 theoretically based on a tape model that incorporates the adhesion and friction forces originating from the van der Waals forces between the submicron-sized spatulae and the substrate (see Figure 6.2), which are controlled by the (macroscopic) actions of the Gecko toes. In order to understand the mechanism introduced in Reference 8, we have to acquaint the Hamaker constant, which is important for understanding dry adhesion.

6.2.4 HAMAKER CONSTANT, ITS ORIGIN AND VALUE

Until know, we kept the microscopic approach; however, the Gecko toe and the sur-face in contact with the Gecko toe are macroscopic objects. So, we have to jump to the macroscopic level, and to ask the question: what interaction force between two macroscopic objects is, namely: the Gecko toe and surface contacting this toe. The second question to be addressed: how this macroscopic interaction arises from microscopic van der Waals forces, addressed in the previous section. So we have to address the following question: what is the macroscopic force acting between two macroscopic bodies interacting via van der Waals forces? The answer to this question depends on the shapes of the interacting bodies and various geometries of the contacting bodies are possible, namely: two flat surfaces, two spheres two cylinders may interact via van der Waals forces. These geometries are addressed in detail in Reference 10. A Gecko toe contacting the solid surface is reasonably treated within the configuration, at which two flat surfaces contact each other, as depicted

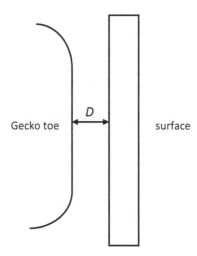

FIGURE 6.5 Gecko toe contacting the surface is reasonably treated in the flat surface/flat surface configuration.

in Figure 6.5. The procedure of integration of van der Walls forces, is cumbersome and it yields finally the following equation for the force per unit area, denoted as $\tilde{F}$:

$$\tilde{F} = -\frac{A}{6\pi D^3},\tag{6.25}$$

where D is separation between surfaces (see Figure 6.5), and A is the so-called Hamaker constant. We already met the Hamaker constant in Section 2.4, when we spoke about the phenomenon of the disjoining pressure. Typical values for the Hamaker constants of condensed phases are $A \cong 10^{-19}$ J for interactions in a vacuum. The values of the adhesion forces for various configurations of contacting bodies are summarized in Reference 10.

It was suggested that the unusual ability of Geckoes to run along vertical forces is to a large extent stipulated by the van der Waals forces. Israelachvili et al. in Reference 8 suggested that the rapid switching between Gecko foot attachment arises from the interplay of the adhesion and friction forces originating from the van der Waals forces between the submicron-sized spatulae and the substrate (see Figure 6.2), which are controlled by the macroscopic actions of the Gecko toes. The pulling force of a spatula along its shaft with an angle θ between 0 and 90° to the substrate has a "normal adhesion force" contribution, produced at the spatula-substrate bifurcation zone and a "lateral friction force" contribution from the part of spatula still in contact with the substrate. High net friction and adhesion forces on the whole Gecko are obtained by rolling down and gripping the toes inward to realize small pulling angles θ between the large number of spatulae in contact with the substrate. To detach, the high adhesion/friction is rapidly reduced to a very low value by rolling the toes upward and backward, which, mediated by the lever function of the setal shaft, peels the spatulae off perpendicularly from the substrates. By these mechanisms, both the adhesion and friction forces of Geckos can be changed over three orders of magnitude, allowing for the swift attachment and detachment during Gecko motion.[8] The effect is amplified by the "contact splitting" phenomenon, discussed in Section 6.1 and developed in References 6–8.

6.2.5 GECKO EFFECT AND DEVELOPMENT OF NOVEL DRY ADHESIVES

"Contact-splitting principle", discussed in Section 6.1, inspired development of novel dry adhesives, based on fiber array surfaces with flat or hemisphere ends.[15] Fiber arrays with different end shapes, such as mush-rooms, asymmetric spatulae, and concave structures, have been reported.[15] In practice, the mushroom-shaped fiber array is the most commonly used.[15] The incorporation of hierarchical structures into fibers has also been explored (compare with the concept of hierarchical reliefs, promoting the Lotus effect, discussed in Sections 2.18 and 2.19).[16] Two- or three-level fiber arrays produce higher adhesion strength than single-level one. Fabrication techniques such as replica molding, 3D direct laser writing, dip transfer processing, fused deposition modeling and digital light processing used for manufacturing Gecko adhesives were covered in Reference 17. These new adhesives enable development and manufacturing of Gecko-inspired robots which may climb and move on wet and humid surfaces.[17]

Bullets

- Geckos are the complex hairy attachment structures among climbing animals that are capable of smart adhesion – the ability to cling to different smooth and rough surfaces and detach at will.
- This ability opens technological pathways for development of novel dry adhesives.
- The classical macroscopic approach to adhesion is supplied by the Johnson–Kendall–Roberts (JKR) theory.
- The pull-off force for a large contact is smaller than the sum of the pull-off forces for many smaller contacts together, covering the same apparent contact area. This concept is known as the "principle of contact splitting".
- From the microscopic point of view, the Gecko adhesion is due to the London (dispersion) forces which are rooted in quantum mechanics.

EXERCISES

1. Spherical body R is in a contact with a plane made of the same material. The surface energy of each body is γ. What is the energy of adhesion between the bodies?

   Solution:

   For two identical surfaces, we calculate: $\hat{\gamma} = \gamma_1 + \gamma_2 - \gamma_{12} = \gamma + \gamma - 0 = 2\gamma$. The adhesion force is given by $F_s = \frac{3}{2}\pi\hat{\gamma}\frac{R_1 R_2}{R_1 + R_2}$. For the contact of spherical bodu with a plane, we assume $R_1 = R; R_2 = \infty$.. Thus, $F_s = \frac{3}{2}\pi\hat{\gamma}\frac{R_1 R_2}{R_1 + R_2} = \frac{3}{2}\pi\hat{\gamma} = 3\pi\gamma$.

2. Two identical spherical balls ($R = 2$ cm) made from mica are in contact. What is the adhesion force acting between the balls? The surface energy of mica equals $\gamma = 80\frac{mJ}{m^2}$

   Solution: $F_s = \frac{3}{4}\pi\hat{\gamma}R = \frac{3}{2}\gamma R = 2.4$ mN.

3. Calculate the total energy of interaction of the given molecule with a diameter d with all other molecules confined within a spherical box L. The intermolecular pair potential is given by $U(r) = -C \frac{x_1 x_2}{r^n}$.

   Hint: The total energy of interaction U_{tot} is given by $U_{tot} = \int_d^L U(r) 4\pi r^2 dr$. Now substitute the intermolecular potential and make carefully the calculations. The result is given by Eq. (6.12).

4. Please explain the notions of the "long-range" and "short-range" forces.

5. Please supply the examples of the short-range and long-range forces.

6. Explain the phenomenon of long-range alignment of magnetic dipoles.

7. Derive the expression for the Coulomb force from the Coulomb potential given by Eq. (6.13).

   Hint: use $F(r) = -\frac{dU}{dr}$.

   Answer: $F(r) = \frac{z_1 z_2 e^2}{4\pi\varepsilon_0 \varepsilon_r r^2}$.

8. Calculate the energy of interaction between neighboring ions of Na^+ and Cl^- in the vacuum. The separation between ions is $r = 0.276\,$nm. Compare this energy to the thermal energy of ions at $T = 300\,$K.

   Hint: For the calculation of the potential energy of interaction between the ions, use Eq. (6.8). $U = -8.36\times10^{-19}\,$J. Thermal energy is estimated as $U_{therm} \cong k_B T$, k_B is the Boltzmann, $k_B = 1.38\times10^{-23}\,\frac{J}{K}$. $U \sim 300 U_{therm}$.

9. What is the physical meaning of the Madelung constant? What are the dimensions of the Madelung constant? Download from the network the values of the Madelung constant obtained for the various ionic crystals (References 10 and 11 are useful for this purpose). What are typical values of the Madelung constant?

10. Calculate the modulus of the molar lattice energy U_{mol} for NaCl crystals.

    Solution: The modulus of the molar lattice energy $|U_{mol}|$ for NaCl crystals is calculated as follows: $|U_{mol}| = N_a U_{tot}$, where $N_a = 6.02\times10^{23}\,$mol^{-1} is the Avogadro number, and U_{tot} is given by Eq. (6.14). $|U_{mol}| \cong 880\,\frac{kJ}{mol}$.

11. Explain the origin of the Keesom dipole-dipole interactions. Why interactions between two dipole molecules should be necessarily average over the possible orientation angles? Check the dimensions in Eq. (6.16).

12. What are the dimensions of the molecular polarizability, defined by Eq. (6.17)?

    Answer: $[\alpha] = \frac{C^2 m}{N} = \frac{Cm^2}{V}$.

13. Derive Eq. (6.18).

    Hint: Simple and elegant derivation of Eq. (6.18) is supplied in Reference 9.

14. Derive Eq. (6.16) from Eq. (6.21) by substitution $\alpha_i = \frac{p_i^2}{3k_B T}$, $i = 1, 2$.

15. Draw van der Waals isotherms (lines of the constant temperatures) emerging from Eq. (6.23), for different temperatures of a gas.

16. Explain how Eq. (6.23) predicts the phase transformations in gases.

17. Explain qualitatively origin of the London dispersion forces.

18. Estimate the London dispersion forces for a number of molecules with Eq. (6.24). Polarizability and ionization potential pf molecules take from the internet databases. Compare the obtained values to the thermal energy of molecules at different temperatures.

REFERENCES

1. Bhushan B. Gecko Effect, in *Encyclopedia of Nanotechnology*; Bhushan B., Ed.; Springer, Dordrecht, 2012.

2. Aowphol A., Yodthong S., Rujirawan A., Kumthorn R., Thirakhupt K. Mitochondrial diversity and phylogeographic patterns of Gekko gecko (Squamata-Gekkonidae) in Thailand. *Asian Herpetological Res.* 2019, **10** (3), 158–169.

3. Hertz H. *Miscellaneous Papers*, London: Macmillan, UK, 1896; p. 146.
4. Johnson K. L., Kendall K., Roberts A. D. Surface energy and the contact of elastic solids. *Proc. R. Soc. London A* 1971, **324** (1558), 301–313.
5. Kroner E., Arzt E. Gecko Adhesion, in Encyclopedia of Nanotechnology; Bhushan B., Ed.; Springer, Dordrecht, 2014.
6. Boesel L. F., Greiner C., Arzt E., del Campo A. Gecko-inspired surfaces: A path to strong and reversible dry adhesives. *Adv. Mater.* 2010, **22**, 2125–2137.
7. Arzt E., Gorb S., Spolenak R. From micro to nano contacts in biological attachment devices. *Proc. Natl. Acad. Sci. USA* 2003, **100**, 10603–10606.
8. Tian Y., Pesika N., Zeng H., Israelachvili J. Adhesion and friction in gecko toe attachment and detachment. *Proc. Natl. Acad. Sci. USA* 2006, **103** (51), 19320–19325.
9. Erbil H. Y. *Solid and Liquid Interfaces*, Blackwell Publishing, Oxford UK, 2006.
10. Israelachvili J. N. *Intermolecular and Surface Forces, 3rd Ed.,* Elsevier, Amsterdam, Netherlands, 2011.
11. Borwein D., Borwein J. M., Taylor K. F. Convergence of lattice sums and Madelung's constant. *J. Math. Phys.* 1985, **26** (11), 2999–3009.
12. Bormashenko E., Gendelman O. On the applicability of the equipartition theorem. *Therm. Phys.* 2010, **14** (3), 855–858.
13. Schroeder D. V. *An Introduction to Thermal Physics*, Addison Wesley, Sam Francisco, CA, USA, 2000.
14. Bormashenko E. Why are the values of the surface tension of most organic liquids similar? *Am. J. Phys.* 2010, **78**, 1309–1311.
15. Zhou M., Pesika N., Zeng H., Israelachvili J. Recent advances in gecko adhesion and friction mechanisms and development of gecko-inspired dry adhesive surfaces. *Friction* 2013, **1**, 114–129.
16. Greiner C., Arzt E., Campo D. A. Hierarchical gecko-like adhesives. *Adv. Mater.* 2009, **21**, 479–482.
17. Sikdar S., Rahman H., Siddaiah A., Menezes P. L. Gecko-inspired adhesive mechanisms and adhesives for robots—A review. *Robotics* 2022, **11** (6), 143.

7 Waves

Now we completely change our strategy in the development of new materials, namely we aim to present novel approaches, resulting in the creation and manufacturing of "metamaterials". What are metamaterials? A metamaterial (from the Greek word μετά *meta*, meaning "beyond" or "after", and the Latin word *materia*, meaning "matter" or "material") is any material engineered to have a property that is rarely observed in naturally occurring materials. Or, in certain cases, a property which is not observed in nature at all. When we develop biomimetic or bioinspired materials, we copy the useful physical properties already observed in nature (such as self-cleaning properties of Lotus leaves or ability of Geckoes to climb along vertical surfaces). We now adopt an opposite approach: we'll try to create a material/metamaterial with a property, which does not occur in nature. As an example, we will discuss metamaterials demonstrating negative refraction index, negative effective density and negative Poisson modulus. These properties are absent in natural materials.

Potential applications of metamaterials are diverse and include super-lenses, medical and electronic devices and even shielding structures from earthquakes. In order to understand how metamaterials are developed, and how do they work, we have to acquaint the phenomenon of wave propagation, in its relation to the physical properties of the medium, conducting the running (traveling) waves. Thus, we start from "oscillations" and "waves".

7.1 OSCILLATIONS AND WAVES

Waves are ubiquitous in nature and engineering and they are familiar to us. We are well aware of ocean and sea waves traveling along the surfaces of large masses of water. Waves transfer sound, when they originate on piano strings or trumpets and reach our ears as sound waves, transferring the beautiful music composed by Bach or Mozart. Waves may appear as traveling waves, which move in some direction, as ocean surf moves toward a beach. And also they may appear as standing waves, like the vibrations of guitar string, in which a wave oscillates in time but its peak amplitude profile does not move in space. Rigorously speaking, wave is a propagation of disturbances from place to place in a regular and organized way. Waves are usually defined as spatially organized oscillations. So, we start from the deep analysis of the phenomenon of oscillations. Let us begin with its simplest manifestation, namely: simple harmonic motion.

7.1.1 SIMPLE HARMONIC MOTION

The well-known examples of the simple harmonic motion, also labeled as "harmonic oscillations" are the motion followed by a mass on the end of the ideal spring (the

DOI: 10.1201/9781003178477-7

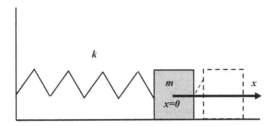

FIGURE 7.1 Scheme of the simple harmonic oscillator is presented. Rigid body m is connected to the massless Hookean spring k (see Eq. (7.1)).

exact meaning of the notion of "ideal spring" will be clarified below) and the motion a pendulum.[1-3]

Consider first the system built of the ideal spring k and rigid mass m, depicted in Figure 7.1. The spring is massless and when deformed by x from the equilibrium position the elastic force on a mass m is described by Hooke's law (see Figure 7.1):

$$F = -kx. \tag{7.1}$$

Such a massless linear spring is called an *ideal spring*. We also adopt that all kinds of friction in the system are zero. Such a system, built of the mass m and ideal spring k, is called a simple harmonic oscillator. The equation describing the motion of the mass m is given by formula (7.2):

$$-kx = ma = m\ddot{x}, \tag{7.2}$$

where $a(t) = \ddot{x}(t) = \frac{d^2 x(t)}{dt^2}$ is the acceleration of mass m. Eq. (7.2) may be rewritten as follows:

$$m\ddot{x} + kx = 0. \tag{7.3}$$

It is convenient to re-shape Eq. (7.3) as follows:

$$m\ddot{x} + kx = 0 \rightarrow \ddot{x} + \frac{k}{m}x = \ddot{x} + \omega^2 x = 0, \tag{7.4}$$

where $\omega^2 = \frac{k}{m}$; $\omega = \sqrt{\frac{k}{m}}$. Eq. (7.4), which is one of the most important in physics and materials science, is called *the equation of a simple harmonic oscillator*. Eq. (7.3) is the homogeneous second-order differential equation. The solution of this equation is given by Eq. (7.5):

$$x(t) = A\sin(\omega t + \varphi), \tag{7.5}$$

where A is called the amplitude of oscillations, and φ is called the phase. The sine function appearing in Eq. (7.5) is dimensionless and moreover it varies between -1 and $+1$; thus, displacement x has the same dimensions as amplitude A, which is defined to be positive and describes the magnitude of the maximum excursion

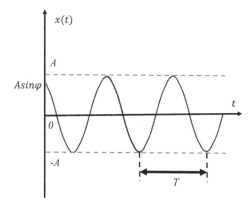

FIGURE 7.2 Harmonic oscillations described by the sine function are illustrated (see Eq. (7.4)). The amplitude A, period T and the phase φ of harmonic oscillations $x(t) = A\sin(\omega t + \varphi)$ are indicated.

away of the mass m from the point of the zero displacement, as illustrated in Figure 7.2.

What is the meaning of constant ω? The constant ω has to do with the repetition time of the oscillatory motion. The sine function repeats itself when the angle increases by 2π rad, (when phase φ is constant), when ωt is increased by 2π. The time necessary for this is defined as a period of oscillations, and labeled T, so that $= 2\pi$. Thus, Eq. (7.6) takes place:

$$\omega = \frac{2\pi}{T}. \tag{7.6}$$

Parameter ω is called the "angular frequency", and it quantifies angular displacement per unit time. Its units are therefore degrees (or radians) per second. The frequency v, in turn, is defined according to Eq. (7.7):

$$v = \frac{1}{T}. \tag{7.7}$$

If the period is measured in seconds, the frequency is measured in s^{-1}. In SI, the unit s^{-1} is labeled the Hertz, after the physicist Heinrich Rudolf Hertz, who first conclusively proved the existence of the electromagnetic waves. Obvious interrelation between the angular frequency ω and frequency v is supplied by Eq. (7.7):

$$\omega = 2\pi v. \tag{7.8}$$

How the constants A (amplitude) and φ appearing in Eq. (7.4) may be calculated? Obviously, these constants could not be derived from the equation of a simple harmonic oscillator (Eq. (7.3)) itself. These constants emerge from the initial conditions to be defined separately, namely: $x(t=0) = x_0; v(t=0) = v_0$.

7.1.2 SIMPLE HARMONIC OSCILLATIONS AND COMPLEX NUMBERS

Now we develop an elegant mathematical technique enabling re-consideration of harmonic oscillations within a very useful and general mathematical framework, namely we'll describe the harmonic oscillations with the complex numbers. The detailed introduction to this technique is presented in the excellent course by Richard Feynman, appearing as Reference 3 in our chapter. For a more close and mathematically rigorous acquaintance with complex numbers, see Reference 4. Complex numbers have real and imaginary parts, and which can be represented on a diagram in which the ordinate represents the imaginary part, and the abscissa represents the real part, as shown in Figure 7.3.

If z is a complex number, we may write it as $z = z_r + jz_i$; $j = \sqrt{-1}$, subscript r means the real part of z, and the subscript i means the imaginary part of z. The real part of the complex number z is denoted $Re(z)$; the imaginary part of the complex number z is denoted $\text{Im}(z)$; thus, $z = z_r + jz_i = Re(z) + jIm(z)$.

Considering Figure 7.3, we may re-shape this representation of a complex number as follows:

$$z = x + jy = re^{j\theta} = r(cos\theta + jsin\theta), \tag{7.9}$$

where r is the absolute value of z (see Figure 7.3 for its geometrical meaning), and θ is the argument of z, $r = \sqrt{x^2 + y^2}$; $tan\theta = \frac{y}{x}$. Eq. (7.9) exploits the very important Euler formula:

$$e^{j\theta} = cos\theta + jsin\theta. \tag{7.10}$$

The squared absolute value of z, denoted r, is calculated as follows:

$$r^2 = x^2 + y^2 = (x + jy)(x - jy) = zz^*, \tag{7.11}$$

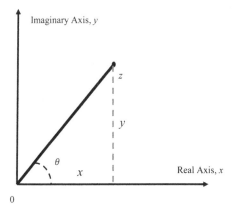

FIGURE 7.3 Geometrical representation of imaginary numbers. $z = x + jy = re^{j\theta} = r(cos\theta + jsin\theta)$. $r = \sqrt{x^2 + y^2}$; $tan\theta = \frac{y}{x}$.

where z^* denotes the complex conjugate of z, and it is obtained by reversing the sign of j in $z = x + jy$. It is easily seen that $Re(z) = \frac{1}{2}(z + z^*)$.

How do we use complex numbers for the analysis of harmonic oscillations? The equation of the harmonic oscillator is $\ddot{x} + \omega^2 x = 0$; the function given by Eq. (7.11) supplies the solution to this equation:

$$x(t) = \hat{C}_1 e^{j\omega t} + \hat{C}_2 e^{-j\omega t}, \tag{7.12}$$

where $\hat{C}_1$ and $\hat{C}_2$ are complex numbers (check this statement). The function $x(t)$ is given now by the complex number; however, we are in the realm of physics, so obviously the position of our oscillator $x(t)$ always has to be a real number. The only way that this can make sense is if choosing physical (i.e., real) initial conditions for the displacement and velocity which give us an answer that stays real all time; otherwise, we've made a mistake somewhere. With the use of Eq. (7.12), we obtain:

$$x(t = 0) = x_0 = \hat{C}_1 + \hat{C}_2 \tag{7.13a}$$

$$v(t = 0) = v_0 = j\omega(\hat{C}_1 - \hat{C}_2). \tag{7.13b}$$

Again, $x(t = 0) = x_0$ and $v(t = 0) = v_0$ are pure real; this is possible only if $\hat{C}_1 + \hat{C}_2$ is real and if $\hat{C}_1 - \hat{C}_2$ is imaginary. This, in turn, means that numbers $\hat{C}_1$ and $\hat{C}_2$ are complex conjugates. Let us represent $\hat{C}_1 = x + jy$, this necessarily implies $\hat{C}_2 = x - jy$. In other words, number $\hat{C}_2$ is complex conjugate to $\hat{C}_1$, i.e., $\hat{C}_2 = \hat{C}_1^*$ takes place. Thus, Eq. (7.12) is re-shaped as follows:

$$x(t) = \hat{C}_1 e^{j\omega t} + \hat{C}_1^* e^{-j\omega t}. \tag{7.14}$$

This might look confusing, since we now seemingly have one arbitrary constant instead of the two, we need. But it is one arbitrary complex number, which is equivalent to two arbitrary reals. In fact, we fulfill the demands of the initial conditions (Eqs. (7.13a,b)), if we assume:

$$\hat{C}_1 = \frac{x_0}{2} - \frac{jv_0}{2\omega}. \tag{7.15}$$

Let us take close look at Eq. (7.14), actually $\hat{C}_1^* e^{-j\omega t}$ is a complex conjugate to $\hat{C}_1 e^{j\omega t}$ (see Exercise (7.19)). Considering $Re(z) = \frac{1}{2}(z + z^*)$ we obtain finally:

$$x(t) = 2Re(\hat{C}_1 e^{j\omega t}). \tag{7.16}$$

Eq. (7.16) has the exact physical sense, because $x(t)$ is represented with Eq. (7.16) by a real number. Consider, however, that we did not just arbitrarily "take the real part" to obtain this result; it emerges out of our initial conditions, given by Eqs. (7.13a,b). It is not correct to solve over the complex numbers and then just ignore half the solution! Anyway, it is much more convenient to use the complex numbers in the problems, related to oscillations and waves and we'll use this technique extensively in our book. For deepening of the understanding of the role of the complex numbers in oscillations and waves theory, see References 3 and 4.

7.1.3 ENERGY IN THE SIMPLE HARMONIC OSCILLATIONS

Total energy of the simple harmonic oscillations E is built of the potential energy of the spring $E_p = \frac{1}{2} kx(t)^2$ and kinetic energy of the mass $E_k = \frac{1}{2} mv(t)^2$ (the spring is assumed to be massless, and its kinetic energy is zero):

$$E = \frac{1}{2} kx(t)^2 + \frac{1}{2} mv(t)^2. \qquad (7.17)$$

Actually, the energy flows between E_k and E_p under simple harmonic oscillations; however, the total energy of the oscillations is conserved (see Exercise (7.10)):

$$E = \frac{1}{2} kA^2 = const. \qquad (7.18)$$

Now consider Eq. (7.4): $x(t) = A\sin(\omega t + \varphi)$. The velocity of the oscillating mass is calculated as follows: $v(t) = \dot{x}(t) = \frac{dx(t)}{dt} = A\omega\cos(\omega t + \varphi)$. Thus, the maximal velocity v_{max} of the oscillating mass is given by $v_{max} = A\omega$. Now consider: $\omega^2 = \frac{k}{m} \rightarrow k = m\omega^2$. Thus, Eq. (7.17) may be rewritten as follows: $E = \frac{1}{2} kA^2 = \frac{1}{2} m\omega^2 A^2 = \frac{1}{2} mv_{max}^2$. It is easily seen now the energy flow occurring between E_k and E_p under simple harmonic oscillations.

7.1.4 THE SIMPLE PENDULUM

One more important particular example of simple harmonic oscillations should be considered, and this is the motion of pendulum. We addressed the idealized pendulum, represented by a point mass m suspended from a massless non-stretchable string l, as shown in Figure 7.4.

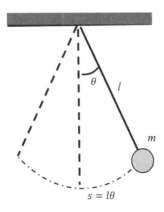

FIGURE 7.4 Simple pendulum is depicted. Point mass m is attached to the massless string l.

The mass moves along the arc of a circle traced out by the end of the taut string (see Figure 7.4). The position s of the mass along the arc of the circle is given by Eq. (7.19) (see Figure 7.4):

$$s = l\theta, \tag{7.19}$$

where angle θ is measured in radians. We omit the details of derivation of the law of the mass m motion, supplied in much detail in Chapter 13 of Reference 2 (which is strongly recommended for the deep study of harmonic oscillations), and supply the Newton second law for a simple pendulum, resulting in Eq. (7.20):

$$ml\frac{d^2\theta}{dt^2} = -mg\sin\theta. \tag{7.20}$$

Cancelling the mass from this equation yields:

$$l\frac{d^2\theta}{dt^2} = -g\sin\theta. \tag{7.21}$$

Regrettably, this differential equation is a non-linear differential equation, the solution of which poses essential mathematical difficulties (for the exact solution of Eq. (7.21), see Reference 5). The solution of Eq. (7.21) becomes simple, if angle θ is small. Indeed, the expansion of the sine function in the Taylor series is given by Eq. (7.22) (see Reference 4):

$$\sin\theta = \theta - \frac{\theta^3}{3!} + \frac{\theta^5}{5!} - \dots \tag{7.22}$$

When angle θ is small (again, it should be stressed that angle θ should be measured in radians), the terms of order θ^3, θ^5 and so forth may be omitted (see Exercise (7.12)). Thus, for small θ, Eq. (7.21) is simplified as follows:

$$l\frac{d^2\theta}{dt^2} = -g\theta. \tag{7.23}$$

Please compare this equation with Eq. (7.3), we must change only the labeling of variables to go from spring to pendulum, according to the following scheme: $x \to \theta$, $k \to g$, $m \to l$. Now we came to the extremely important concept of the materials science: *similar differential equations yield similar physical behavior of physical systems*. Indeed, the solution of the differential Eq. (7.23) is given by Eq. (7.24):

$$\theta(t) = \theta_0 \sin(\omega t + \delta); \ \omega = \sqrt{\frac{g}{l}}, \tag{7.24}$$

where θ_0 and δ are the amplitude and initial phase of the oscillations to be established from the initial conditions, namely $\theta(t=0) = \theta_0$; $\omega(t=0) = \frac{d\theta(t)}{dt}(t=0) = \omega_0$.

Compare this solution with that supplied by Eq. (7.4). The period of frequency of the pendulum's oscillations are given by Eq. (7.25):

$$T = 2\pi \sqrt{\frac{l}{g}}; \upsilon = \frac{1}{2\pi}\sqrt{\frac{g}{l}}. \tag{7.25}$$

7.1.5 Damped Harmonic Motion

Almost all real physical systems, including oscillating systems, are affected by friction, which was completely neglected in our previous treatment. Actually, friction is a very complicated phenomenon, as discussed in detail in Reference 6, and hardly may be "enveloped" within a single formula. In order to make the problem understandable we adopt the simplified model, in which friction is introduced as the "drag force", described by Eq. (5.2), when we treated the "shark-skin" effect. For a purpose of brevity, we rewrite Eq. (5.2), quantifying the friction force denoted f_{fr} as follows:

$$f_{fr} = -b\dot{x} = -b\frac{dx}{dt}, \tag{7.26}$$

where b is the damping coefficient or damping parameter (the example of the damped system is depicted in Figure 7.5). The friction force described by Eq. (7.26) is a dissipative force and it converts irreversibly the mechanical energy into heat.[5] Again, it should be emphasized that Eq. (7.26) represents a very specific kind of friction, occurring under the viscous mechanism of energy dissipation.

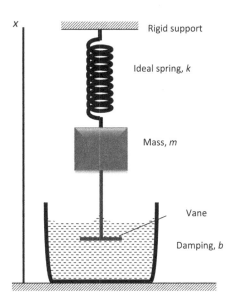

FIGURE 7.5 Damped oscillator is depicted. The viscous friction force is given by $f_{fr} = -b\dot{x} = -b\frac{dx}{dt}$; b is the damping coefficient.

If the friction is described by Eq. (7.26), the Newton second law for the oscillations of the damped system appears as follows:

$$-kx - b\dot{x} = m\ddot{x}. \tag{7.27}$$

It is useful to rewrite Eq. (7.27), as it is set below:

$$\ddot{x} + \frac{b}{m}\dot{x} + \frac{k}{m}x = 0 \tag{7.28}$$

We introduce now the following notation, which will be used across the entire book: $\frac{b}{2m} = \beta$; $\frac{k}{m} = \omega_0^2$; β is called the "damping factor" ($[\beta] = s^{-1}$) and ω_0 is the angular frequency in the absence of drag.[2,7] Thus, Eq. (7.28) is re-shaped as follows:

$$\ddot{x} + 2\beta\dot{x} + \omega_0^2 x = 0. \tag{7.29}$$

The solution of this equation (which is called "the equation of damped oscillations") is given by Eq. (7.30):

$$x(t) = Ae^{-\beta t}\sin(\omega' t + \varphi), \tag{7.30}$$

where $\omega' = \sqrt{\omega_0^2 - \beta^2}$ is the modified angular frequency (A and φ should be established from the initial conditions). It is also useful to supply the solution of Eq. (7.29) obtained with the use of complex numbers:

$$x(t) = \hat{C}_+ e^{-j\omega_+ t} + \hat{C}_- e^{-j\omega_- t}, \tag{7.31}$$

where $\omega_\pm = -j\beta \pm \sqrt{\omega_0^2 - \beta^2}$, $\hat{C}_+$ and $\hat{C}_-$ are the complex constants to be established from the initial conditions. The falling (decreasing) function in Eq. (7.30) is a kind of envelope (shown by dashed lines, in Figure 7.6), that modulates what would otherwise be simple harmonic oscillation; this causes the amplitude of the motion to decrease as times goes on (indeed, $Ae^{-\beta t}$ may be reasonably interpreted as the amplitude of oscillations which is decreased exponentially with time). The mass eventually comes to rest at equilibrium point (consider Eq. (7.30) and the definition of the damping factor $\beta = \frac{b}{2m}$; it turns out the argument of exponential is directly proportional to b, that is to the value of drag force).[2]

The difference between the angular frequency of the damped oscillations $\omega' = \sqrt{\omega_0^2 - \beta^2}$ and that of non-damped ones $\omega_0 = \sqrt{\frac{k}{m}}$ is intuitively not so clear, but it should be emphasized, due to its crucial importance for understanding of damped oscillations. When $\omega' = \sqrt{\omega_0^2 - \beta^2} = 0$ we say that the system is critically damped; this takes place when the condition:

$$b_c = \sqrt{4km} = 2m\omega_0. \tag{7.32}$$

is fulfilled. When Eq. (7.32) takes place, the system is critically damped.[2,7] The system is underdamped for $b < b_c$ (this situation is depicted in Figure 7.6); in this

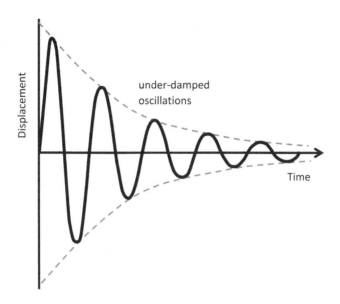

FIGURE 7.6   The solution of the equation of damped oscillations (Eq. (7.29)) is presented. The dashed envelope line depicts the decreasing amplitude of oscillations. The case of light damping (or underdamping) is depicted.

case, the system exhibits oscillatory behavior – albeit with exponentially decreasing amplitude, as illustrated in Figure 7.6.

When $b > b_c$, the system is overdamped; in other words, the system comes to rest at equilibrium without oscillations; it crosses the equilibrium point at most once following an initial displacement.[2,7]

7.1.6   FORCED OSCILLATIONS

Consider a rigid subject to an ideal (Hookean) spring force (see Eq. (7.1)), a drag force proportional to a speed (see Eq. (7.26)) and an external harmonic driving force, described by Eq. (7.33).

$$F(t) = F_0 cos(\omega t), \tag{7.33}$$

where F_0 is an amplitude of the driving force and ω is the angular frequency of its temporal change, which is called the driving frequency in this case. Motion of the mass is now described by Eq. (7.33):

$$-kx - b\dot{x} + F_0 cos(\omega t) = m\ddot{x}. \tag{7.34}$$

It is convenient to re-shape Eq. (7.34) as follows (as it was already carried out for the damped oscillations (compare with Eq. (7.29)):

$$\ddot{x} + 2\beta\dot{x} + \omega_0^2 x = f_0 cos(\omega t), \tag{7.35}$$

where $\omega_0 = \sqrt{\frac{k}{m}}$, $\beta = \frac{b}{2m}$ and $f_0 = \frac{F_0}{m}$. The solution of Eq. (7.35) is given by Eq. (7.36):

$$x(t) = Ae^{-\beta t}\sin\left(\omega' t + \varphi\right) + \frac{f_0}{\sqrt{\left(\omega_0^2 - \omega^2\right)^2 + 4\beta^2\omega^2}}\cos\left[\omega t + arctg\left(\frac{2\beta\omega}{\omega_0^2 - \omega^2}\right)\right], \quad (7.36)$$

where $\omega' = \sqrt{\omega_0^2 - \beta^2}$ is the modified angular frequency (A and φ should be established from the initial conditions). The first term of the solution, given by Eq. (7.36) represent exponential effects due to the friction (compare to Eq. (7.30)), and it is important only at the first stage of oscillations. These effects emerging from friction will die with time (the characteristic time of decay of the effects due to friction $\tau \sim \frac{1}{\beta}$). The second term represents harmonic motion with the frequency of the driving force ω and amplitude $\tilde{A}$ given by Eq. (7.37):

$$\tilde{A} = \frac{f_0}{\sqrt{\left(\omega_0^2 - \omega^2\right)^2 + 4\beta^2\omega^2}}. \quad (7.37)$$

In other words, after the transient time $\sim \frac{1}{\beta}$, the motion of the mass is simple harmonic motion with the frequency of driving force ω and amplitude $\tilde{A}$ given by Eq. (7.37), as illustrated in Figure 7.7. This is intuitively clear: imagine, for example, a spring with a mass on its end suspended from a harmonically driven hand. In time $\tau \sim \frac{1}{\beta}$ the mass will move with the frequency of the hand motion, even if the motion of the mass and your hand are not in phase.

The amplitude, given by Eq. (7.37) has the remarkable property of resonance, it comes to maxim under the certain value of the frequency of driving force ω. In order to find this frequency we have to find the maximum of $\tilde{A}(\omega)$, or in other words the minimum of denominator of Eq. (7.37), for this purpose let us establish the minimum of the expression: $(\omega_0^2 - \omega^2)^2 + 4\beta^2\omega^2$; let us calculate the derivative of this expression and equal it to zero:

$$-4\left(\omega_0^2 - \omega^2\right)\omega + 8\beta^2\omega = 0. \quad (7.38)$$

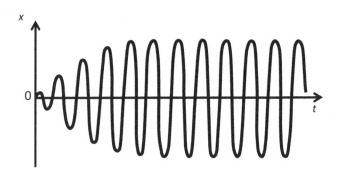

FIGURE 7.7 Oscillations forced by the external sinusoidal force $F(t) = F_0\cos(\omega t)$ are depicted. After the transient time $\sim \frac{1}{\beta}$, the motion of the mass is simple harmonic motion with the frequency of driving force ω and amplitude $\tilde{A}$ given by Eq. (7.37).

Eq. (7.38) has three solutions: (i) $\omega=0$ corresponds to the maximum of denominator; (ii) $\omega=-\sqrt{\omega_0^2-2\beta^2}$ has no physical sense; (iii) $\omega=\sqrt{\omega_0^2-2\beta^2}$ corresponds to the minimum of denominator and maximum of the amplitude $\tilde{A}$ supplied by Eq. (7.37). This is the resonance frequency of the system $\omega_{res}=\sqrt{\omega_0^2-2\beta^2}$. Substitution of $\omega_{res}=\sqrt{\omega_0^2-2\beta^2}$ into Eq. (7.37) yields:

$$\tilde{A}_{res}=\frac{f_0}{2\beta\sqrt{\omega_0^2-\beta^2}}. \tag{7.39}$$

We recognize from Eq. (7.39) that if the drag force is zero ($\beta=0$), the amplitude $\tilde{A}_{res}$ diverges and goes to infinity. Moreover, if the drag force is zero ($\beta=0$), we calculate $\omega_{res}=\sqrt{\omega_0^2-2\beta^2}=\omega_0$. The sharpness of the resonance peak depends on interrelation between the damping factor $\beta=\frac{b}{2m}$ and frequency $\omega_0=\sqrt{\frac{k}{m}}$, as illustrated in Figure 7.8.

The use of complex numbers for the analysis of the forced equations yields ($z=x+jy$ is a complex number):

$$\ddot{z}+2\beta\dot{z}+\omega_0^2 z=f_0 e^{j\omega t}. \tag{7.40}$$

For a steady-state solution of Eq. (7.40), we try $z=z_0 e^{j\omega t}$ and calculate:

$$z_0=\frac{f_0}{\omega_0^2-\omega^2+2j\beta\omega}. \tag{7.41}$$

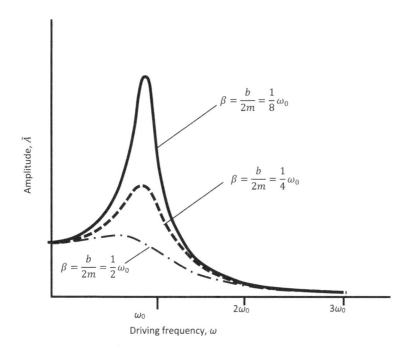

FIGURE 7.8 The dependencies of amplitude of the forced oscillations $\tilde{A}$ on the driving frequency ω (also called resonance curves) for different damping factors $\beta=\frac{b}{2m}$ are depicted.

For the real modulus of the complex amplitude, we obtain:

$$|z_0| = \sqrt{z_0 z_0^*} = \frac{f_0}{\sqrt{\left(\omega_0^2 - \omega^2\right)^2 + 4\beta^2 \omega^2}}, \tag{7.42}$$

which should be compared to Eq. (7.37). We again recognize that the use of complex numbers is an elegant way for solution of the problems related to harmonic oscillations.

7.1.7 WAVES: THE WAVE EQUATION

Waves are spatially organized oscillations; let us start from the waves propagating in the elastic string. Consider thin homogeneous string with a constant linear density depicted in Figure 7.9. The linear density $\mu = \frac{m}{L}$ (m is the mass of the segment L) is supposed to be constant along the string ($[\mu] = \frac{kg}{m}$). We adopt that the string has been initially displaced in the transverse direction (direction y in Figure 7.9).

We also adopt that distortion of the string is small, meaning that i) the tension T is the same throughout strings (this will be demonstrated below, this is not so clear, due to the string has the mass); ii) angles θ are small (see Figure 7.9). Let us write the Newton second law for the infinitesimal element of the string dm, as shown in Figure 7.9:

$$T_2 \sin\theta_2 - T_1 \sin\theta_1 = a_y dm, \tag{7.43a}$$

$$T_2 \cos\theta_2 - T_1 \cos\theta_1 = a_x dm. \tag{7.43b}$$

where a_x and a_y are the horizontal and vertical components of the acceleration of the element dm correspondingly. Angles θ_1 and θ_2 are supposed to be small; thus, $\cos\theta_2 \cong \cos\theta_1 \cong 1$. We also assume that the motion of the mass dm is pure transversal,

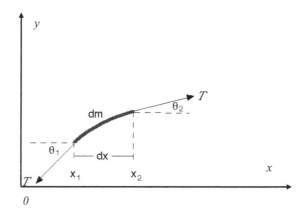

FIGURE 7.9 Element dm of the string is displaced transversally. The length of the element is dx. The angles θ_1 and θ_2 are supposed to be small; thus, the magnitudes of the tension T are equal.

so $a_x = 0$; thus we obtain from Eq. (7.43b): $T_2 \cong T_1 = T$. Assuming also $sin\theta \cong tg\theta \cong \theta$ and considering $a_y = \frac{\partial^2 y(x,t)}{\partial t^2}$ we derive from Eqs. (7.41a) to (7.41b):

$$Ttg\theta = a_y dm = \frac{\partial^2 y(x,t)}{\partial t^2} dm. \qquad (7.44)$$

Now we take into account $\mu = \frac{m}{L} \to m = \mu L \to dm = \mu dx$, and $tg\theta = \frac{\Delta y}{\Delta x}$ obtain:

$$T\frac{\Delta y}{\Delta x} = \frac{\partial^2 y(x,t)}{\partial t^2} \mu dx. \qquad (7.45)$$

Eq. (7.45) is easily re-formulated in the language of partial derivatives:

$$\frac{\partial^2 y(x,t)}{\partial t^2} = \frac{T}{\mu}\frac{\partial^2 y(x,t)}{\partial x^2}. \qquad (7.46)$$

We identify in Eq. (7.46) the coefficient $\frac{T}{\mu}$ relating the partial derivatives one to another. What is the physical meaning of this coefficient? Let us analyze its dimensions: $\left[\frac{T}{\mu}\right] = \frac{kg \times m}{s^2} \times \frac{m}{kg} = \frac{m^2}{s^2} = [v^2]$. We established that the dimensions of the coefficient $\frac{T}{\mu}$ coincide with those of the squared velocity. Thus, Eq. (7.46) is re-shaped as follows:

$$\frac{\partial^2 y(x,t)}{\partial t^2} = v^2 \frac{\partial^2 y(x,t)}{\partial x^2}, \qquad (7.47)$$

where $v = \sqrt{\frac{T}{\mu}}$. The exact meaning of this velocity will be clarified below. Eq. (7.47) is called the "wave equation", and it is of crucial importance for this part of the course, devoted to metamaterials. Its solution is given by Eq. (7.48) (see Exercise (7.22)):

$$y(x,t) = f(x \pm vt). \qquad (7.48)$$

Functions $y(x,t) = f(x \pm vt)$ represent shapes that move to the right with velocity v, or to the left with velocity $-v$ as illustrated in Figure 7.10. Thus, indeed, the physical meaning of the velocity $v = \sqrt{\frac{T}{\mu}}$ coincides with the velocity of propagation of the wave along axis x. The wave described by Eq. (7.48) is also called the *traveling wave* (to be distinguished from the *standing wave*, also known as a stationary wave, which waves that oscillates in time but whose peak amplitude profile does not move in space; standing waves occur inside a resonator due to interference between waves reflected back and forth at the resonator's resonant frequency[1,2]). Actually, v appearing in Eqs. (7.47) and (7.48) is the *phase velocity of the traveling wave*, as it is will be demonstrated below.

An important particular case of the solution of the wave equation is supplied by harmonic functions (sine or cosine). Consider the sinusoidal wave that propagates with velocity v. In such a wave we have a doubled oscillations occurring in space in time, namely during the single period of oscillations T the sine wave moves (propagates) to the distance λ, which is called the wavelength (the shape of the wave repeats itself over time T), as depicted in Figure 7.11.

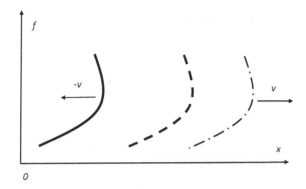

FIGURE 7.10 The curve is solution of the wave equation represented by the functions $f(x \pm vt)$. As the time goes on, the curve $f(x - vt)$ moves to the right with speed v, whereas the curve $f(x + vt)$ moves to the left with speed $-v$.

It is convenient to present the sinusoidal solution of the wave equation with Eq. (7.49):

$$y(x,t) = A\sin\frac{2\pi}{\lambda}(x - vt) = A\sin 2\pi\left(\frac{x}{\lambda} - \frac{vt}{\lambda}\right), \tag{7.49}$$

where A is the amplitude of the wave; considering $\lambda = vT$ yields Eq. (7.50)

$$y(x,t) = A\sin 2\pi\left(\frac{x}{\lambda} - \frac{t}{T}\right). \tag{7.50}$$

The double space/time nature of oscillations occurring within a traveling wave are clearly recognized from Eq. (7.50). Now we introduce the notion of the wave-number defines as $k = \frac{2\pi}{\lambda}$, $[k] = m^{-1}$. Considering this definition and taking into account $\omega = \frac{2\pi}{T}$ we obtain finally Eq. (7.51):

$$y(x,t) = A\sin(kx - \omega t). \tag{7.51}$$

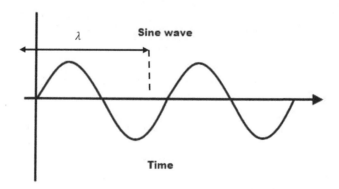

FIGURE 7.11 Wavelength λ of the sinusoidal wave is depicted.

TABLE 7.1

Parameters Quantifying Temporal and Spatial Periodicity
Inherent for Harmonic Waves

Temporal Periodicity	Spatial Periodicity
Period, T	Wavelength, λ
Angular frequency, $\omega = \frac{2\pi}{T}$	Wavenumber, $k = \frac{2\pi}{\lambda}$

Eq. (7.51) also expresses the double spatial/temporal nature of periodicity occurring within the sinusoidal harmonic waves. The physical meaning of the wavenumber $k = \frac{2\pi}{\lambda}$ is a "spatial angular frequency" of the wave; it shows, as a matter of fact, how many complete waves may be inserted with 2π meters of the given direction. The useful analogy between temporal and spatial periodicity inherent for harmonic waves is supplied in Table 7.1. The temporal and spatial periodicities in the harmonic waves are coupled via the velocity of propagation of the wave; consider that $\lambda = vT$ takes place. Considering $k = \frac{2\pi}{\lambda}$ and $\omega = \frac{2\pi}{T}$ and substitution into $\lambda = vT$ immediately yields Eq. (7.51):

$$\omega = vk, \qquad (7.52)$$

establishing interrelation between the angular frequency and wavenumber of sinusoidal wave, which also illustrates the coupling between temporal and spatial periodicities in the sinusoidal waves. Eq. (7.52) is also called the *dispersion relation* and it is very important for our future analysis of metamaterials. Looking ahead, let us mention that in the general case the dispersion relation in the medium appears as follows:

$$\omega(k) = v(k)k. \qquad (7.53)$$

Eq. (7.53) implies that the velocity of the wave propagation depends on the wavenumber. More elegant representation of the sinusoidal waves is supplied by the complex numbers. The solution of the wave equation in this case is supplied by Eq. (7.54):

$$y(x,t) = Ae^{j(kx-\omega t)}. \qquad (7.54)$$

The waves described by Eq. (7.51) (or alternatively Eq. (7.54)) are called monochromatic (one frequency) harmonic (sinusoidal) traveling waves.

7.1.8 Longitudinal Waves

Until now, we addressed the transverse waves, motion in which all points on a wave oscillate along paths at right angles to the direction of the wave's advance. Surface ripples on water, and electromagnetic waves (to be addressed further in detail) are examples of transverse waves. Now we consider the longitudinal waves, which are waves in which the vibration of the medium is parallel to the direction the wave

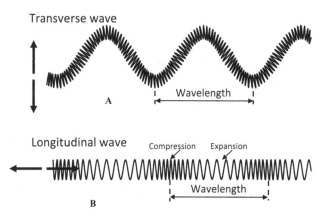

FIGURE 7.12   Transverse (A) vs. longitudinal (B) waves are depicted.

travels and displacement of the medium is in the same (or opposite) direction of the wave propagation. The illustrative comparison between propagation of transversal and longitudinal waves is supplied in Figure 7.12.

Consider now the tube, filled with the gas shown in Figure 7.13. The gas is driven by the piston, which moves harmonically according to the law $\xi(t)=\xi_0 cos(\omega t)$, where ξ is the horizontal displacement of the piston (see Figure 7.13).

The particles of the gas, which are close to the piston move according to the same law, namely $\xi(t)=\xi_0 cos(\omega t)$, however the particles of the gas located at a distance x from the piston move with the delay $\tau = \frac{x}{v}$, where v is the velocity of the wave propagation. The law of the motion of these particles is described by Eq. (7.55):

$$\xi(x,t)=\xi_0 cos\omega(t-\tau)=\xi_0 cos\omega\left(t-\frac{x}{v}\right)=\xi_0 cos\left(\omega t-\omega\frac{x}{v}\right). \qquad (7.55)$$

Obviously, the function $\xi(x,t)=\xi_0 cos\omega\left(t+\frac{x}{v}\right)$ describes the wave that propagates in the negative direction of axis x. Actually, Eq. (7.55) represents the equation of a *plane wave*. In physics, a *plane wave* is a special case of wave or field: a physical quantity whose value, at any moment, is constant through any plane that is perpendicular to a fixed direction in space, in our case the displacements of the particles constituting

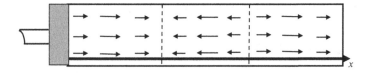

FIGURE 7.13   Longitudinal waves propagating within a tube filled with a gas are depicted. The gas is driven by the piston performing harmonic oscillations according to the $\xi(t)=\xi_0 cos(\omega t)$. Arrows depict the displacement of the particle constituting the gas, which are also involved into the harmonic motion.

the gas are constant within a plane normal to axis x. More accurately speaking, Eq. (7.55) represents the plane harmonic wave. In the general three-dimensional (3D) case the equation of the monochromatic plane wave is written as follows:

$$\xi(\vec{r},t)=\xi_0 cos\left(\omega t-\vec{k}\cdot\vec{r}\right), \tag{7.56}$$

and when we use the complex numbers Eq. (7.56) is rewritten as follows:

$$\xi(\vec{r},t)=\xi_0 e^{j\left(\omega t-\vec{k}\cdot\vec{r}\right)}, \tag{7.57}$$

where $\xi(\vec{r},t)$ denotes the propagating physical value, $\vec{k}$ and ω are the wave-vector (to be defined rigorously below) and the angular frequency of the wave (again the double/coupled periodicities are inherent for the longitudinal plane waves). The wave-vector $\vec{k}$ appears in Eqs. (7.56) and (7.57) as vector value; let us clarify its exact meaning. Consider a 3D plane wave, propagating with velocity v, depicted in Figure 7.14. Assume that oscillations occurring in the plane passing through the origin are harmonic, namely $\xi=\xi_0 cos\omega t$ takes place in this plane (see Figure 7.14). Vector normal to this plane is denoted $\vec{n}$.

Vibrations in the plane distanced by l from this plane are described by the formula $\xi=\xi_0 cos\omega\left(t-\frac{l}{c}\right)$; radius-vector of the points belonging to this plane is denoted $\vec{r}$ (see Figure 7.14). Obvious geometric interrelation $l=\vec{n}\cdot\vec{r}$ takes place; thus, we derive:

$$\xi=\xi_0 cos\omega\left(t-\frac{\vec{n}\cdot\vec{r}}{v}\right)=\xi_0 cos\left(\omega t-\frac{\omega}{v}\vec{n}\cdot\vec{r}\right). \tag{7.58}$$

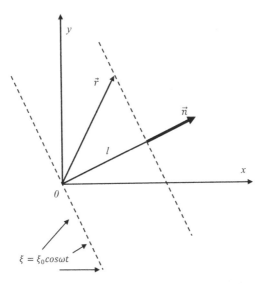

FIGURE 7.14 Plane wave propagating in the direction $\vec{n}$ is depicted. Oscillations occurring in the plane passing through the origin are harmonic, namely $\xi=\xi_0 cos\omega t$ takes place in this plane.

Now consider the dispersion relation given by Eq. (7.52), namely: $\frac{\omega}{v} = k$. We derive:

$$\xi = \xi_0 \cos(\omega t - k\vec{n} \cdot \vec{r}). \tag{7.59}$$

We define wave-vector $\vec{k}$ with Eq. (7.60):

$$\vec{k} = k\vec{n}. \tag{7.60}$$

Thus, we re-shape Eq. (7.58) as follows: $\xi(\vec{r}, t) = \xi_0 \cos(\omega t - \vec{k} \cdot \vec{r})$ (compare with Eq. (7.56)). The magnitude of the wave-vector $k = |\vec{k}|$ plays exactly the same role that k played in the 1-D case. That is, k is the wavenumber. It equals 2π times the number of wavelengths that fit into a unit length. The modulus of the wave-vector $\vec{k}$ equals to the wavenumber: $|\vec{k}| = k = \frac{2\pi}{\lambda}$. The waves, described by Eqs. (7.56)–(7.59) are classified as monochromatic (one frequency), longitudinal waves. The 3D wave equation, describing the 3D propagation of the wave appears as follows:

$$\frac{\partial^2 \xi(x, y, zt)}{\partial t^2} = v^2 \left(\frac{\partial^2 \xi(x, y, z, t)}{\partial x^2} + \frac{\partial^2 \xi(x, y, z, t)}{\partial y^2} + \frac{\partial^2 \xi(x, y, z, t)}{\partial z^2} \right), \tag{7.61}$$

where v is the velocity of the wave propagation. What is the value of velocity of propagation of longitudinal waves and how it is related to the physical properties of the material? This problem is of primary importance for materials science. Recall, that the velocity of the transversal waves is given by $v = \sqrt{\frac{T}{\mu}}$, where T is the tension of the string and μ is the linear mass given by $\mu = \frac{m}{L}$ (m is the mass of the segment L). The wave velocity generally may be re-formulated as $v = \sqrt{\frac{resoring\ force\ factor}{mass\ factor}}$ (see Reference 2). The same relation remains true also for longitudinal waves, namely, the velocity of propagation of longitudinal waves in the elastic medium with the modulus Young E is given by Eq. (7.62):

$$v = \sqrt{\frac{E}{\rho}}, \tag{7.62}$$

where ρ is the density of the medium. The velocity of propagation of the transversal waves in the elastic medium in given, in turn, by Eq. (7.63):

$$v = \sqrt{\frac{G}{\rho}}, \tag{7.63}$$

where G is the shear modulus of the elastic medium. For the exact definitions of the Young and shear moduli see Reference 8; Reference 8 is strongly recommended for the first acquaintance with the materials science.

7.1.9 PHASE AND GROUP VELOCITIES OF THE WAVE

Now we introduce the very important concept, which plays a crucial role in the understanding how the metamaterials work; this is a concept of the group velocity of the wave. We already acquainted with the velocity of the wave $v = \frac{\omega}{k}$ (see Eq. (7.52)), and expressed with Eqs. (7.62) and (7.63). This wave velocity is actually the *phase velocity* of the wave. Now we are going to acquaint ourselves with the group velocity of the wave. For this purpose, consider two harmonic sinusoidal waves $y_1(x, t)$ and $y_2(x,t)$ which meet in a certain point in the space. The waves are described with the following equations (compare with Eq. (7.51)):

$$y_1(x,t) = A\sin(\omega_1 t - k_1 x) \tag{7.64a}$$

$$y_2(x,t) = A\sin(\omega_2 t - k_2 x), \tag{7.64b}$$

where ω_1 and ω_2 are the angular frequencies of the waves, and k_1, k_2 are their wavenumbers. For the sake of simplicity, we assume that the amplitudes of the waves A are the same. We also assume that superposition of waves is linear; thus, their sum is simply expressed with Eq. (7.65):

$$y(x,t) = y_1(x,t) + y_2(x,t). \tag{7.65}$$

Substitution of Eqs. (7.64a,b) into Eq. (7.65) yields:

$$y(x,t) = A\{\sin(\omega_1 t - k_1 x) + \sin(\omega_2 t - k_2 x)\}. \tag{7.66}$$

Consider $\sin\alpha + \sin\beta = 2\sin\frac{\alpha+\beta}{2}\cos\frac{\alpha-\beta}{2}$; thus, we derive:

$$y = 2A\sin\frac{(\omega_1+\omega_2)t - (k_1+k_2)x}{2}\cos\frac{(\omega_1-\omega_2)t-(k_1-k_2)x}{2}. \tag{7.67}$$

Now we introduce the "averaged" (denoted ω_{av} and k_{av}) and "modulated" (denoted ω_{mod} and k_{mod}) frequencies and wavenumbers with Eqs. (7.67a,b):

$$\omega_{av} = \frac{\omega_1+\omega_2}{2}; \ k_{av} = \frac{k_1+k_2}{2}. \tag{7.68a}$$

$$\omega_{mod} = \frac{\omega_1-\omega_2}{2}; \ k_{mod} = \frac{k_1-k_2}{2}. \tag{7.68b}$$

Thus, Eq. (7.67) is re-shaped as follows:

$$y(x,t) = 2A\cos(\omega_{mod}t - k_{mod}x)\sin(\omega_{av}t - k_{av}x). \tag{7.69}$$

Eqs. (7.68) and (7.69) state that the sum of original two waves can be re-formulated as a product of two *traveling waves*; the wavelength and the frequency of the second traveling wave, which is the sine term in Eq. (7.69) are given by the average of corresponding quantities for the original two waves. Eq. (7.69) represents the high frequency and high wavenumber wave (ω_{av}, k_{av}) enveloped with the low-frequency wave (ω_{mod}, k_{mod}). The envelope formed by the first (cosine) wave is the dashed curve in Figure 7.15. This is not so obvious; thus, we'll suggest one more assumption, which will make the situation more transparent and understandable.

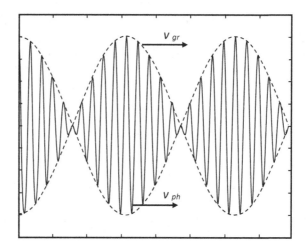

FIGURE 7.15 Modulated traveling wave is depicted. Phase (v_{ph}) and group (v_{gr}) velocities of the wave are illustrated. The dashed line shows the low-frequency cosine envelope encompassing high-frequency sinusoidal oscillations (see Eq. (7.69)). The group velocity v_{gr} is the velocity of motion of the entire envelope; v_{gr} quantifies the velocity of transfer of information.

We adopt that angular frequencies ω_1 and ω_2 are close one to another, and the same is true for the wavenumbers k_1 and k_2. Rigorously speaking we assume that Eq. (7.69) holds:

$$\omega_{mod} \ll \omega_{av}; k_{mod} \ll k_{av}. \tag{7.70}$$

Let us re-shape Eq. (7.69) as follows:

$$y(x,t) = A_{mod}(x,t) \sin(\omega_{av}t - k_{av}x), \tag{7.71}$$

where $A_{mod}(x,t)$ is given by Eq. (7.72):

$$A_{mod}(x,t) = 2A\cos(\omega_{mod}t - k_{mod}x). \tag{7.72}$$

Eqs. (7.71) and (7.72) describe the amplitude-modulated almost sinusoidal traveling wave. The amplitude $A_{mod}(x,t) = 2A\cos(\omega_{mod}t - k_{mod}x)$ varies slowly in time and space, when compared with the rapid processes quantified with ω_{av}, k_{av} (recall that the wavenumber is a "space frequency" of the traveling wave). Thus, we obtain the rapidly changing process enveloped with a slowly changing amplitude, as shown in Figure 7.15. In this situation, we say that the high-frequency process is modulated with the low-frequency one, or, in other words: we recognize the *amplitude modulation* in Figure 7.15. Consider that the picture presented in Figure 7.15 is not a static one; the entire shape is moving along axis x. Now we ask the very important question: at what velocity does the modulation propagate? The question is re-formulated as follows: at what velocity a given modulation wave crest (i.e., a point at which $A_{mod}(x,t) = 2 A$ travels)? To follow a given constant value (such as a crest) of the modulation amplitude $A_{mod}(x,t)$ we need to maintain a constant value to its argument (see Eq. (7.72)):

$$\omega_{mod} t - \kappa_{mod} x = const \tag{7.73}$$

Differentiation of Eq. (7.73) yields immediately Eq. (7.74):

$$\omega_{mod} dt - \kappa_{mod} dx = 0 \tag{7.74}$$

Thus, we derive for the modulation velocity, which we denote v_{gr}:

$$v_{gr} = \frac{dx}{dt} = \frac{\omega_{mod}}{k_{mod}} \tag{7.75}$$

Substitution $\omega_{mod} = \frac{\omega_1 - \omega_2}{2}; k_{mod} = \frac{k_1 - k_2}{2}$ (see Eq. (7.68b)) yields:

$$v_{gr} = \frac{dx}{dt} = \frac{\omega_1 - \omega_2}{k_1 - k_2} = \frac{\Delta\omega}{\Delta k} \tag{7.76}$$

Now we assume: $\Delta\omega \to 0; \Delta k \to 0$; thus, in the limiting case we obtain:

$$v_{gr} = \frac{d\omega(k)}{dk} \tag{7.77}$$

The velocity supplied by Eq. (7.77) is called the group velocity of the wave. The group velocity of a wave is the velocity with which the overall envelope shape of the wave's amplitudes, known as the modulation or envelope of the wave, (see Figure 7.15) propagates through space. In the practice of modern wave communication, the envelope, shown with a dashed line in Figure 7.15, transfers the signal. Thus, we see that a "signal" consisting of a wave crest of the modulation propagates not with a velocity $v = \frac{\omega}{k}$ (see Eq. (7.52)) but at the group velocity $v_{gr} = \frac{d\omega(k)}{dk}$. The velocity emerging from Eq. (7.52) we'll call in our future analysis "the phase velocity of the wave" and denote $v_{ph} = \frac{\omega}{k}$. This is the velocity of the monochromatic, harmonic/sinusoidal wave. From the point of view of sending information, these waves are not useful. They are the same throughout time and space. Something must therefore be modulated, such as frequency or amplitude, in order to convey information, as illustrated in Figure 7.15. The information/signal is conveyed with the group phase $v_{gr} = \frac{d\omega(k)}{dk}$; this is the velocity of motion of envelope/information, as shown in Figure 7.15 with a dashed line. The interrelation between the group and phase velocities of the waves will be clarified below. Consider the definition of the group velocity $v_{gr} = \frac{d\omega(k)}{dk}$ and $\omega = v_{ph}k$. Thus, we obtain:

$$v_{gr} = \frac{d(v_{ph}k)}{dk} = v_{ph} + k\frac{dv_{ph}(k)}{dk}. \tag{7.78}$$

As an example, we consider propagation of electromagnetic waves in ionosphere.[1] The dispersion relation for this kind of waves is given by Eq. (7.79):

$$\omega^2 = \omega_p^2 + c^2 k^2, \tag{7.79}$$

where $\omega_p = const$ is the so-called plasma frequency, which will be addressed further in much detail (see Section 8.9) and c is the velocity of light in vacuum. Let us calculate the phase and group velocities for this kind of wave. The phase velocity is calculated as follows:

$$v_{ph} = \frac{\omega}{k} = \frac{\sqrt{\omega_p^2 + c^2 k^2}}{k}. \tag{7.80}$$

The group velocity is derived with Eq. (7.81):

$$v_{gr} = \frac{d\omega(k)}{dk} = \frac{c^2 k}{\sqrt{\omega_p^2 + c^2 k^2}} = \frac{c^2}{v_{ph}}. \tag{7.81}$$

Eq. (7.80) leads to the paradoxical conclusion, indeed: $v_{ph} = \frac{\sqrt{\omega_p^2 + c^2 k^2}}{k} = \sqrt{\frac{\omega_p^2}{k^2} + c^2}$. Thus, we conclude that for electromagnetic waves in ionosphere $v_{ph} > c$ is true. However, $v_{gr} = \frac{c^2}{v_{ph}} = \frac{c}{v_{ph}} c < c$; thus, the group velocity is always less than c. Recall that the signal is transferred with the group velocity of the wave; hence, a signal cannot be transmitted with a velocity greater than c, which is the velocity of light in vacuum.

7.1.10 ELECTROMAGNETIC WAVES

One of the most important kinds of transversal waves is electromagnetic waves. Electromagnetic waves are created as a result of vibrations between an electric and a magnetic field. Electromagnetic waves differ from mechanical waves in that they do not require a medium to propagate. This means that electromagnetic waves can travel not only through gases, liquids and solid materials, but also through the vacuum of space. The existence of electromagnetic waves is predicted by the Maxwell equations. The macroscopic Maxwell equations are supplied below:

$$\nabla \cdot \vec{D} = \rho \tag{7.82a}$$

$$\nabla \times \vec{H} = \vec{j} + \frac{\partial \vec{D}}{\partial t} \tag{7.82b}$$

$$\nabla \cdot \vec{B} = 0 \tag{7.82c}$$

$$\nabla \times \vec{E} = -\frac{\partial \vec{B}}{\partial t}, \tag{7.82d}$$

where $\vec{E}$ is the electric field, $\vec{D}$ is the electrical displacement field (also called electrical induction), $\vec{B}$ is the magnetic field, $\vec{H}$ is the magnetic field intensity (also called magnetic field strength), ρ is the electric charge density and $\vec{j}$ is the electric current density, the $\nabla \cdot$ symbol (pronounced "del dot") denotes the divergence operator, $\nabla \times$ (pronounced "del cross") denotes the curl operator. Electric displacement

$\vec{D}$ is a vector that represents that aspect of an electric field associated solely with the presence of separated free electric charges, purposely excluding the contribution of any electric charges bound together in neutral atoms or molecules. Vector of magnetic field intensity/strength $\vec{H}$, in turn, represents the part of the magnetic field in a material that arises from an external current and is not intrinsic to the material itself. Actually, Eq. (7.82a) represents the Gauss law for electrostatics, Eq. (7.82b) expresses Ampère's circuital law (with Maxwell's addition of the displacement current $\frac{\partial \vec{D}}{\partial t}$), Eq. (7.82c) is the Gauss law for magnetism, reflecting the absence of magnetic charges, and Eq. (7.82d) is Faraday's law of induction. For a deep understanding of the Maxwell equations, which is crucially necessary for understanding of electronic metamaterials, see References 9–12. For the deep understanding of the divergence and curl operators, see Reference 13. It should be emphasized that Eqs. (7.82a–d) are supplied in the SI system of units.

The system of Maxwell equations is not complete until the interrelation between vectors $\vec{D}$ and $\vec{E}$ and also $\vec{B}$ and $\vec{H}$ is established. We also need to prescribe the relation between the current density and the electric field $\vec{E}$. These interrelations are supplied by Eqs. (7.83a–c):

$$\vec{D} = \varepsilon_0 \varepsilon_r \vec{E} = \varepsilon \vec{E} \tag{7.83a}$$

$$\vec{B} = \mu_0 \mu_r \vec{H} = \mu \vec{H} \tag{7.83b}$$

$$\vec{j} = \sigma \vec{E}, \tag{7.83c}$$

where $\varepsilon_0 \cong 8.85 \times 10^{-12} \frac{F}{m}$ (Farads per meter) is the absolute dielectric vacuum permittivity and $\mu_0 = 4\pi \times 10^{-7} \frac{H}{m}$ (Henry per meter) is the magnetic permeability of vacuum; the constants ε_0 and μ_0 emerge in the SI system of units; ε_r and μ_r are relative electric and magnetic permittivity of the medium (ε_r is also often called the dielectric constant of the medium), σ is the electric conductivity of the medium; $\mu = \mu_0 \mu_r$ is the magnetic permeability of the medium, $\varepsilon = \varepsilon_0 \varepsilon_r$ is the dielectric permittivity of substance (in vacuum $\varepsilon_r = 1$; $\mu_r = 1$). The velocity of light in vacuum is given by $c = \frac{1}{\sqrt{\varepsilon_0 \mu_0}}$.

Eqs. (7.83a–c) include the electric and magnetic characteristics of the specific medium, and this is the point, where the materials science engineer starts his work. Electromagnetic metamaterials are artificial materials characterized with unusual values of ε and μ, which do not occur naturally. How these materials became possible will be discussed later. It should be emphasized that Eqs. (7.83a–c) are true in the isotropic media; the extension of these equations to anisotropic media is found in References 3 and 12.

Now we are ready to formulate equations governing propagation of electromagnetic waves in media. These equations (Eqs. (7.84a,b)) emerge from formulas (7.81) and (7.82); for rigorous derivations of these equations, see References 10 and 11.

$$\frac{\partial^2 \vec{E}(x,y,z,t)}{\partial t^2} = \frac{c^2}{\varepsilon_r \mu_r} \left(\frac{\partial^2 \vec{E}(x,y,z,t)}{\partial x^2} + \frac{\partial^2 \vec{E}(x,y,z,t)}{\partial y^2} + \frac{\partial^2 \vec{E}(x,y,z,t)}{\partial z^2} \right) \tag{7.84a}$$

$$\frac{\partial^2 \vec{H}(x,y,z,t)}{\partial t^2} = \frac{c^2}{\varepsilon_r \mu_r} \left(\frac{\partial^2 \vec{H}(x,y,z,t)}{\partial x^2} + \frac{\partial^2 \vec{H}(x,y,z,t)}{\partial y^2} + \frac{\partial^2 \vec{H}(x,y,z,t)}{\partial z^2} \right). \qquad (7.84b)$$

Eqs. (7.84a,b) may be rewritten in a more compact form:

$$\frac{\partial^2 \vec{E}(x,y,zt)}{\partial t^2} = \frac{c^2}{\varepsilon_r \mu_r} \nabla^2 \vec{E}(x,y,z,t) \qquad (7.85a)$$

$$\frac{\partial^2 \vec{H}(x,y,zt)}{\partial t^2} = \frac{c^2}{\varepsilon_r \mu_r} \nabla^2 \vec{H}(x,y,z,t), \qquad (7.85b)$$

where ∇^2 is the Laplace operator (also denoted Δ), which a second-order differential operator, defined as follows $\nabla^2 f(x_1 \dots x_n) = \Delta f(x_1 \dots x_n) = \sum_{i=1}^{n} \frac{\partial^2 f(x_1 \dots x_n)}{\partial x_i^2}$ (for details, see Reference 13). Compare these equations with Eq. (7.47) and Eq. (7.61). It is immediately recognized from Eqs. (7.84) to (7.85a,b) that they describe waves in which vectors of electric and magnetic fields $\vec{E}$ and $\vec{H}$ oscillate being coupled in space and time. Consider the electromagnetic wave propagating in the z-direction, depicted in Figure 7.16. Direction of the electric field $\vec{E}$ and magnetic field $\vec{H}$ is perpendicular to the direction of wave propagation, coinciding with the direction of wave-vector $\vec{k}$ (see Figure 7.16). Neither the electric field nor the magnetic field has a component in the direction of wave propagation. It is also generally true that the electric and the magnetic field are perpendicular to each other, as shown in Figure 7.16. Thus, the condition $\vec{E} \cdot \vec{H} = 0$ takes place. It also should be emphasized that vectors $\vec{E}$, $\vec{H}$ and $\vec{k}$ form the right triplet, in other words they form the right-handed vector system (the system of unit-vectors $\vec{i}$, $\vec{j}$ and $\vec{k}$ is called right-handed if $\vec{i} \times \vec{j} = \vec{k}$). Looking ahead, we note that vectors $\vec{E}$, $\vec{H}$ and $\vec{k}$ form the right triplet in the "normal", "conventional"

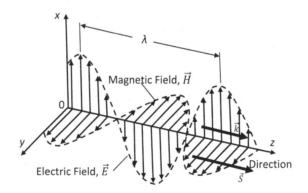

FIGURE 7.16 Scheme of traveling electromagnetic wave is depicted; the wavelength of the wave is λ. Electric $\vec{E}$ and magnetic $\vec{H}$ fields are coupled within the electromagnetic field. Wave-vector $\vec{k}$ indicates the direction of the wave propagation. Vectors $\vec{E}$, $\vec{H}$ and $\vec{k}$ form the right triplet. $\vec{S}$ is the Poynting vector. Electromagnetic waves differ from mechanical waves in that they do not require a medium to propagate. This means that electromagnetic waves can travel not only through air and solid materials, but also through the vacuum of space.

materials and also in vacuum. In metamaterials $\vec{E}$, $\vec{H}$ and $\vec{k}$ form the left-handed vector system.

It also follows from Eqs. (7.84a,b) that the phase velocity of the electromagnetic waves is given by Eq. (7.86):

$$v_{ph} = \frac{c}{\sqrt{\varepsilon_r \mu_r}}. \tag{7.86}$$

Just Eq. (7.86) opened the way to the development of metamaterials; controlling dielectric and magnetic properties of the medium, enables control of the phase velocity of the electromagnetic wave propagation; this, in turn, provides the possibility of the developments of the materials with the controlled refraction index, as it will be discussed in much detail in the following chapters. In vacuum $\varepsilon_r = 1, \mu_r = 1$ and hence $v_{ph} = c \cong 2.99792 \times 10^8 \, \frac{m}{s}$, which is the velocity of light in vacuum. In vacuum, the wave equations describing propagation of electromagnetic waves appear as follows:

$$\frac{\partial^2 \vec{E}(x,y,zt)}{\partial t^2} = c^2 \nabla^2 \vec{E}(x,y,z,t) \tag{7.87a}$$

$$\frac{\partial^2 \vec{H}(x,y,zt)}{\partial t^2} = c^2 \nabla^2 \vec{H}(x,y,z,t). \tag{7.87b}$$

One of the possible solutions of Eqs. (7.84) and (7.85) and Eq. (7.87) is supplied by Eqs. (7.88a,b)

$$\vec{E}(\vec{r},t) = \vec{E}_0 e^{j(\omega t - \vec{k} \cdot \vec{r})} \tag{7.88a}$$

$$\vec{H}(\vec{r},t) = \vec{H}_0 e^{j(\omega t - \vec{k} \cdot \vec{r})}. \tag{7.88b}$$

Eqs. (7.88a,b) describe 3D *plane monochromatic electromagnetic wave*; vector $\vec{k}$ indicates the direction of propagation of the wave, as depicted schematically in Figure 7.17. The wave-front of the wave is perpendicular to the vector $\vec{k}$ (the wave-front of a wave is the set (locus) of all points having the same phase); in our case, it is the set of points in which vectors $\vec{E}(\vec{r},t) = \vec{E}_0 e^{j(\omega t - \vec{k} \cdot \vec{r})}$ and $\vec{H}(\vec{r},t) = \vec{H}_0 e^{j(\omega t - \vec{k} \cdot \vec{r})}$ has the same phase.

The phase velocity of the 3D monochromatic electromagnetic wave is given by:

$$v_{ph} = \frac{\omega}{k} = \frac{\omega}{|\vec{k}|} = \frac{\omega}{\sqrt{k_x^2 + k_y^2 + k_z^2}}. \tag{7.89}$$

7.1.11 Energy of Electromagnetic Field—Poynting Vector

Electromagnetic waves transfer energy, let us calculate the energy transported by the plane electromagnetic wave, described by Eqs. (7.88a,b). The density of electromagnetic field energy w is given by (see Exercise (7.33)):

$$w = w_E + w_H = \frac{\varepsilon_0 \varepsilon_r E^2}{2} + \frac{\mu_0 \mu_r H^2}{2}, \tag{7.90}$$

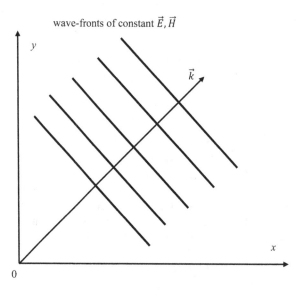

FIGURE 7.17 Propagation of the plane electromagnetic wave is depicted schematically. Wave-fronts of constant $\vec{E}, \vec{H}$ are shown. Wave-vector $\vec{k}$ normal to the wave-fronts indicates the direction of propagation of the wave.

where w_E is the density of electric field energy and w_H is the density of the magnetic field energy with the dimensions: $[w_E]=[w_H]=\frac{J}{m^3}$. It should be emphasized that Eq. (7.90) is true in the medium with zero dispersion, i.e., $\varepsilon \neq \varepsilon(\varepsilon,k)$, $\mu \neq \mu(\omega,k)$ is assumed. Substitution of Eqs. (7.88a,b) into Eqs. (7.84a,b) yields:

$$E_0\sqrt{\varepsilon_0\varepsilon_r} = H_0\sqrt{\mu_0\mu_r}. \tag{7.91}$$

Electric and magnetic fields oscillate within the electromagnetic wave exactly in the same phase; hence, Eq. (7.91) also takes place for the instantaneous values of the electric and magnetic field:

$$E\sqrt{\varepsilon_0\varepsilon_r} = H\sqrt{\mu_0\mu_r}. \tag{7.92}$$

Eqs. (7.90) and (7.92) immediately yield: $w_E = w_H$; thus, we derive:

$$w = 2w_E = \varepsilon_0\varepsilon_r E^2. \tag{7.93}$$

Exploiting Eq. (7.92) yields:

$$w = \sqrt{\varepsilon_0\varepsilon_r\mu_0\mu_r}\, EH. \tag{7.94}$$

Considering Eq. (7.86) $v_{ph} = \frac{c}{\sqrt{\varepsilon_r\mu_r}}$ and $c = \frac{1}{\sqrt{\varepsilon_0\mu_0}}$ gives rise to Eq. (7.95):

$$w = \frac{1}{v_{ph}} EH. \tag{7.95}$$

The energy flux transferred by the electromagnetic wave (denoted S) is supplied by $S = v_{ph}w$. Involving Eq. (7.95) yields:

$$S = EH. \tag{7.96}$$

The energy flux has a certain direction; thus, it is more reasonable to quantify it with the vector of energy flux, denoted $\vec{S}$ and expressed with Eq. (7.97):

$$\vec{S} = \vec{E} \times \vec{H}. \tag{7.97}$$

This vector is called the Poynting vector.[10,11] In "normal", natural materials its direction coincides with the direction of the wave-vector $\vec{k}$, as shown in Figure 7.16. Thus, vectors $\vec{E}$, $\vec{H}$ and $\vec{S}$ form the right triplet (see Figure 7.16). Looking ahead, we note that in a very paradoxical way for the metamaterials it is not true, i.e., vectors $\vec{E}$, $\vec{H}$ and $\vec{S}$ form the left triplet.

Bullets

- Wave is a propagation of disturbances from place to place in a regular and organized way. Waves are spatially organized oscillations.
- The equation of the harmonic oscillator is $\ddot{x} + \omega^2 x = 0$, $\omega = \sqrt{\frac{k}{m}}$. Its solution is given by $x(t) = A sin(\omega t + \varphi)$ or alternatively: $x(t) = 2Re(\hat{C}_1 e^{j\omega t})$.
- The equation of the damped harmonic oscillator is $\ddot{x} + 2\beta\dot{x} + \omega_0^2 x = 0$, where β is the damping factor. Its solution is given by $x(t) = Ae^{-\beta t} sin(\omega' t + \varphi)$, where $\omega' = \sqrt{\omega_0^2 - \beta^2}$.
- The equation of forced oscillations is $\ddot{x} + 2\beta\dot{x} + \omega_0^2 x = f_0 cos(\omega t)$, the amplitude of oscillations is $\tilde{A} = \dfrac{f_0}{\sqrt{\left(\omega_0^2 - \omega^2\right)^2 + 4\beta^2\omega^2}}$. The amplitude of the forced oscillations grows, when the resonance conditions are fulfilled.
- Waves are classified as transversal and longitudinal. Electromagnetic waves are transversal, the sound waves may be transversal and also longitudinal.
- The equation of transversal waves propagating in the string is $\frac{\partial^2 y(x,t)}{\partial t^2} = v^2 \frac{\partial^2 y(x,t)}{\partial x^2}$, $v = \sqrt{\frac{T}{\mu}}$, where T and μ are the tension and linear tension of the string; the sinusoidal solution of this equation is $y(x,t) = A sin(kx - \omega t)$, or alternatively $y(x,t) = Ae^{j(kx-\omega t)}$, where k and ω are the wavenumber and angular frequency of the traveling wave. The dispersion relation $\omega = vk$ takes place.
- Propagating longitudinal plane wave is described by the equation: $(\vec{r},t) = \xi_0 e^{j(\omega t - \vec{k}\cdot\vec{r})}$, where $\xi(\vec{r},t)$ denotes the propagating physical value, $\vec{k}$ and ω are the wave-vector and the angular frequency of the wave.
- The modulated traveling wave is described by $y(x,t) = A_{mod}(x,t) sin(\omega_{av}t - k_{av}x)$; $A_{mod}(x,t) = 2A cos(\omega_{mod}t - k_{mod}x)$.
- The phase velocity of the wave is $v_{ph} = \frac{\omega}{k}$; the group velocity of the wave is $v_{gr} = \frac{d\omega(k)}{dk}$.
- Electromagnetic waves propagating in the medium are described by wave equations: $\frac{\partial^2 \vec{E}(x,y,zt)}{\partial t^2} = \frac{c^2}{\varepsilon_r \mu_r} \nabla^2 \vec{E}(x,y,z,t)$; $\frac{\partial^2 \vec{H}(x,y,zt)}{\partial t^2} = \frac{c^2}{\varepsilon_r \mu_r} \nabla^2 \vec{H}(x,y,z,t)$, where $\vec{E}$ and $\vec{H}$ are vectors of the electric and magnetic field; ε_r and μ_r are

relative electric and magnetic permittivity of the medium, c is the velocity of the light in vacuum. The sinusoidal solutions of these equations are
$\vec{E}(\vec{r},t) = \vec{E}_0 e^{j(\omega t - \vec{k} \cdot \vec{r})}$, $\vec{H}(\vec{r},t) = \vec{H}_0 e^{j(\omega t - \vec{k} \cdot \vec{r})}$, representing plane monochromatic electromagnetic waves.

- The phase velocity of the electromagnetic waves is $v_{ph} = \frac{c}{\sqrt{\varepsilon_r \mu_r}}$.
- The vector of energy flux $\vec{S}$, transferred by the electromagnetic wave (also
  called the Poynting vector) is given by $\vec{S} = \vec{E} \times \vec{H}$.

EXERCISES

1. Derive Eq. (7.3), i.e., the equation of motion of the simple harmonic oscillator.
2. Check the dimensions in the equation: $\omega^2 = \frac{k}{m}$. What are the dimensions
   of ω?
   Answer: $[\omega] = s^{-1}$.
3. What is the physical meaning of the frequency v and angular frequency ω of
   harmonic oscillations?
4. Calculate the velocity of the simple harmonic oscillator.
   Hint: Use Eq. (7.4).
   Answer: $v = \frac{dx}{dt} = A\omega \cos(\omega t + \varphi)$.
5. Calculate the acceleration of the simple harmonic oscillator.
   Answer: $a = \frac{dv}{dt} = -A\omega^2 \sin(\omega t + \varphi) = -\omega^2 x(t)$.
6. Consider the system presented in the scheme below. The system is built of
   the ideal, massless Hookean spring k and mass m.

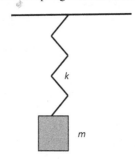

What is the frequency of harmonic oscillations in the discussed system?
Hint: Consider the axis y which origin coincides with the equilibrium
position of the mass m.
Answer: The motion of a mass m on the end of the vertical massless
Hookean spring has exactly the same frequency as it does on the end of a
horizontal spring: $\omega = \sqrt{\frac{k}{m}}$. The only change is that the displacement from
the new equilibrium position.

7. Demonstrate that the function $x = \hat{C}_1 e^{j\omega t} + \hat{C}_2 e^{-j\omega t}$ supplies the solution to the
   equation of the harmonic oscillator: $\ddot{x} + \omega^2 x = 0$.
8. Prove that the real part of the complex number z is given by $Re(z) = \frac{1}{2}(z + z^*)$.
9. Demonstrate that $\hat{C}_1^* e^{-j\omega t}$ is a complex conjugate to $\hat{C}_1 e^{j\omega t}$, whatever is a complex number $\hat{C}_1$.
   Solution: Assume: $\hat{C}_1 = x + jy$, thus: $\hat{C}_1 e^{j\omega t} = (x + jy)(\cos\omega t + j\sin\omega t) =$
   $x\cos\omega t - y\sin\omega t + j(x\sin\omega t + y\cos\omega t)$.
   In turn, $\hat{C}_1^* e^{-j\omega t} = (x - jy)(\cos\omega t - j\sin\omega t) = x\cos\omega t - y\sin\omega t - j(x\sin\omega t +$
   $y\cos\omega t)$; thus, we demonstrated that $\hat{C}_1^* e^{-j\omega t}$ is a complex conjugate to $\hat{C}_1 e^{j\omega t}$.

10. Demonstrate that the total energy of the simple harmonic oscillations is conserved.

Solution: The total energy of the simple harmonic oscillations E is given by Eq. (7.16): $E = \frac{1}{2}kx(t)^2 + \frac{1}{2}mv(t)^2$; the displacement of the mass, in turn, is given by Eq. (7.4): $x(t) = A\sin(\omega t + \varphi)$. The velocity of the oscillating mass is calculated as follows: $v(t) = \dot{x}(t) = \frac{dx(t)}{dt} = A\omega\cos(\omega t + \varphi)$. Thus, the total energy is given by $E = \frac{1}{2}kA^2\sin^2(\omega t + \varphi) + \frac{1}{2}m\omega^2 A^2\cos^2(\omega t + \varphi)$. Now consider: $\omega^2 = \frac{k}{m} \rightarrow m = \frac{k}{\omega^2}$; thus, we derive: $E = \frac{1}{2}kA^2\sin^2(\omega t + \varphi) + \frac{1}{2}\frac{k}{\omega^2}$
$\omega^2 A^2\cos^2(\omega t + \varphi) = \frac{1}{2}kA^2\left[\sin^2(\omega t + \varphi) + \cos^2(\omega t + \varphi)\right] = \frac{1}{2}kA^2$.

11. Derive Eq. (7.20) (see Chapter 13 of Reference 2 for the self-control).

12. For small angles $\sin\theta \cong \theta$ may be assumed. Calculate the difference between θ and $\sin\theta$ for: a) $\theta = 0.2\ rad$, b) $\theta = 0.1\ rad$.

Answer: (a) the difference about 1%; (b) the difference is about 0.1%.

13. Check the dimensions in Eq. (7.25): $= 2\pi\sqrt{\frac{l}{g}}$; $\upsilon = \frac{1}{2\pi}\sqrt{\frac{g}{l}}$.

14. What physical force works as a "restoring force" which is responsible for oscillations of the pendulum, depicted in Figure 7.4.

Answer: The restoring force in this case is gravity (for details, see Reference 2).

15. What are the dimensions of the damping coefficient (damping parameter (appearing in Eq. (7.26) quantifying the friction force: $f_{fr} = -b\dot{x}$.

Answer: $[b] = \frac{kg}{s}$.

16. What are the dimensions of the "damping factor" β appearing in Eq. (7.28)?

Answer: $[\beta] = s^{-1}$.

17. Check by substitution that Eqs. (7.30) and (7.31) represent the solution of the equation of damped oscillations (Eq. (7.29)).

18. Derive the condition of critical damping: $b_c = \sqrt{4km} = 2m\omega_0$. Check the dimensions in this equation.

19. Explain qualitatively the phenomenon of critical damping.

20. Derive the physically meaningful minimum of the minimum of the expression $(\omega_0^2 - \omega^2)^2 + 4\beta^2\omega^2$ appearing in the expression defining the amplitude of forced oscillations given by Eq. (7.3).

21. Use the complex numbers for the solution of the equation of forced oscillations $\ddot{z} + 2\beta\dot{z} + \omega_0^2 z = f_0 e^{j\omega t}$ and calculate the complex amplitude of the steady-state oscillations z_0.

Hint: use $z = z_0 e^{j\omega t}$.

Answer: $z_0 = \frac{f_0}{\omega_0^2 - \omega^2 + 2j\beta\omega}$.

22. Calculate the real modulus of the complex amplitude of the forced oscillations given by Eq. (7.41).

23. Check that the function $y(x,t) = f(x \pm vt)$ supply the solution to the wave equation: $\frac{\partial^2 y(x,t)}{\partial t^2} = v^2\frac{\partial^2 y(x,t)}{\partial x^2}$.

Solution: Denote $f(x \pm vt)$ as $f(u)$, where $u = x \pm vt$; thus, $\frac{\partial u}{\partial x} = 1$; $\frac{\partial u}{\partial t} = \pm v$. Exploit the chain rule: $\frac{\partial f}{\partial x} = \frac{\partial f}{\partial u}\frac{\partial u}{\partial x}$ and correspondingly: $\frac{\partial f}{\partial t} = \frac{\partial f}{\partial u}\frac{\partial u}{\partial t}$. Thus, we obtain $\frac{\partial f}{\partial x} = \frac{\partial f}{\partial u}$ and $\frac{\partial^2 f}{\partial x^2} = \frac{\partial^2 f}{\partial u^2}$; $\frac{\partial f}{\partial t} = \pm v\frac{\partial f}{\partial u}$ and $\frac{\partial^2 f}{\partial t^2} = v^2\frac{\partial^2 f}{\partial u^2}$. Substituting into the wave equation yields: $\frac{\partial^2 f}{\partial x^2} - \frac{1}{v^2}\frac{\partial^2 f}{\partial t^2} = \frac{\partial^2 f}{\partial u^2} - v^2\frac{1}{v^2}\frac{\partial^2 f}{\partial u^2} = 0$.

24. Derive the dispersion relation $\omega = vk$ for harmonic waves.

25. Demonstrate that that the complex function $y(x,t) = Ae^{j(kx - \omega t)}$ supplies the solution to the wave equation: $\frac{\partial^2 y(x,t)}{\partial t^2} = v^2\frac{\partial^2 y(x,t)}{\partial x^2}$.

26. Calculate the velocity of the particles in the longitudinal plane wave.
    Hint: Exploit Eq. (7.55).
    Solution: $v(x,t) = \frac{d\xi(x,t)}{dt} = -\omega\xi_0 \sin\omega\left(t - \frac{x}{c}\right)$.

27. Define the wave-vector $\vec{k}$. What are the direction and modulus of the wave-vector?

28. Check the dimensions in Eqs. (7.62) and (7.63).

29. Clarify the notion of the monochromatic wave. Why monochromatic waves are useless for the transport of information?

30. Derive Eq. (7.69) describing superposition of two harmonic/sinusoidal waves.

31. Derive Eq. (7.77) defining the group velocity of the wave: $v_{gr} = \frac{d\omega(k)}{dk}$.

32. The dispersion relation of deep water gravity waves is given by $\omega = \sqrt{gk}$, where g is the gravity acceleration. Calculate the phase and group velocities for these waves. Establish the interrelation between the phase and group velocities.
    Solution: $v_{ph} = \frac{\omega}{k} = \frac{\sqrt{gk}}{k} = \sqrt{\frac{g}{k}}$. $v_{gr} = \frac{d\omega(k)}{dk} = \frac{1}{2}\sqrt{\frac{g}{k}}$. $v_{ph} = \frac{1}{2}v_{gr}$.

33. The group velocity of the waves is given by $v_{ph} = \alpha k$, $\alpha = const$. Establish the group velocity of the waves.
    Solution: $v_{gr} = \frac{d\omega(k)}{dk}$; $\omega = v_{ph}k = \alpha k^2$; $v_{gr} = \frac{\partial\omega(k)}{\partial k} = 2\alpha k = 2v_{ph}$.

34. Demonstrate using the model of parallel-plates capacitor that the density of the electric field is given by $w_E = \frac{\varepsilon_0\varepsilon E^2}{2}$ (see Eq. (7.90)).
    Solution: The energy of the parallel-plates capacitor is given by $W = \frac{CU^2}{2}$, where C is the capacity and U is voltage; the capacity, in turn, is given by $= \frac{\varepsilon_0\varepsilon S}{d}$, where S is the area of the plates, and d is the distance between the plates (see Reference 9). Thus, the energy of the capacitor is expressed as: $W = \frac{\varepsilon_0\varepsilon S U^2}{2d} = \frac{\varepsilon_0\varepsilon}{2}\left(\frac{U}{d}\right)^2 Sd$. Now consider $E = \frac{U}{d}$; $V = Sd$, where E is the electrical field within the capacitor, and V is the volume of the capacitor. Hence, we obtain $W = \frac{\varepsilon_0\varepsilon}{2}E^2 V$; the volume density of the energy of the electric field within the capacitor $w_E = \frac{W}{V} = \frac{\varepsilon_0\varepsilon}{2}E^2$; the dimensions of the volume density of the energy of the electric field are $[w_E] = \frac{J}{m^3}$.

35. Demonstrate Eq. (7.91).

36. Derive Eq. (7.95).

37. What is the physical meaning of the Poynting vector $\vec{S}$? What are the dimensions of the Poynting vector?
    Solution: $[S] = \frac{J}{m^2 s}$.

38. Check that vectors $\vec{E}$, $\vec{H}$ and $\vec{S}$ form the right triplet ($\vec{S}$ is the Poynting vector).

REFERENCES

1. Crawford F. S. Jr. *Waves, Berkeley Physics Course*, McGraw-Hill, Berkeley, CA, USA, 1968; Vol. 3.

2. Fishbane P. M., Gasiorowicz S., Thornton S. T. Physics for Scientists and Engineers, Chapter 14, *Waves*, Prentice Hall, Upper Saddle River, NJ, USA, 1993; Vol. 1.

3. Feynman R. P., Leighton R. B., Sands M. *The Feynman Lectures on Physics*, Addison-Wesley Pub. Co, Reading, MA, 1963–1965.

4. Arfken G. B., Weber H. J. *Mathematical Methods for Physicists, 5th Ed.*, A Harcourt Science and Technology Company, San Diego, USA, 2001.

5. Landau L. D., Lifshitz E. *Mechanics, Volume 1 of the Course of the Theoretical Physics, 3rd Ed.*, Butterworth-Heinemann, Oxford, UK, 2000.

6. Nosonovsky M., Mortazavi V. *Friction-Induced Vibrations and Self-Organization: Mechanics and Non-Equilibrium Thermodynamics of Sliding Contact*, CRC Press, Boca Raton, FL, USA, 2014.

7. Savelyev I. V., Leib G. *Physics - A General Course*, MIR Publishers, Moscow, USSR, 1981.

8. Callister W. D. Jr. *Materials Science and Engineering. An Introduction, 6th Ed.*, John Wiley & Sons, Hoboken, NJ, USA, 2003.

9. Fishbane P. M., Gasiorowicz S., Thornton S. T. *Physics for Scientists and Engineers, Vol. 2, Chapter 35, Maxwell's Equations and Electromagnetic Waves*, Prentice Hall, Upper Saddle River, NJ, USA, 1993.

10. Purcell E. M., Morin D. J. *Electricity and Magnetism, 3rd Ed.*, Chapter 9, Cambridge University Press, Cambridge, UK, 2013.

11. Panofsky W., Phillips M.. *Classical Electricity and Magnetism, 2nd Ed.*, Dover Publications, Mineola, NY, USA, 2005.

12. Landau L. D., Lifshitz E. M., *Electrodynamics of Continuous Media.* Volume 8 in Course of Theoretical Physics, 2nd Ed., Pergamon Press, Oxford, 1984.

13. Arfken G. B., Weber H. J. *Mathematical Methods for Physicists*, 5th Ed., Harcourt/Academic Press, San Diego, CA, USA, 2001.

8 Refraction Index and Its Origin

We are already acquainted with electromagnetic waves introduced in the previous chapter as coupled transverse waves of electric and magnetic fields. Transverse traveling electromagnetic waves permeate the entire Universe. It was thought for a long time that the fact that light could travel through space necessarily means that space is not empty but filled with the "luminiferous ether". It was emphasized in the previous chapter that light/electromagnetic wave does not need a "material" for propagation but could propagate through vacuum with the speed $c \cong 3.0 \times 10^8 \frac{m}{s}$ (this speed being a fundamental physical constant plays a crucial role in special relativity). Now we study what is observed, when light encounters on its pathway a material/medium; in other words, we study interaction of the traveling electromagnetic wave with a matter. The diversity of physical events occurring when electromagnetic wave is a material is amazing and it is definitely impossible to cover this diversity of physical effects within a single chapter. We pose a modest aim of understanding the physical nature of the refraction index. The attained understanding will pave a way for the development of metamaterials, enabling a controlled propagation of light in media.

8.1 THE INTERACTION OF LIGHT WITH MATERIALS

Interaction of electromagnetic waves with materials depends crucially on the wavelength/frequency of the wave; in vacuum, the interrelation between the frequency υ and wavelength of electromagnetic wave λ is given by Eq. (8.1) (see Eq. (7.52) in the previous chapter):

$$\upsilon = \frac{c}{\lambda}. \tag{8.1}$$

Frequencies and wavelengths of electromagnetic waves cover a very broad range of magnitudes. The electromagnetic spectrum (part of which is depicted with Figure 8.1), covers electromagnetic waves with frequencies ranging from below one hertz to above 10^{25} hertz, corresponding to wavelengths from thousands of kilometers down to a fraction of the size of an atomic nucleus. The range of wavelengths of radiation is vast, the section of the electromagnetic spectrum shown with Figure 8.1 spans 15 orders of magnitude 10^{-13} m $< \lambda < 10^2$ m. The shortest wavelengths appearing in Figure 8.1 (10^{-13} m $< \lambda < 10^{-11}$ m) correspond to Gamma rays, which have the smallest wavelengths and the most energy of any wave in the electromagnetic spectrum. They are produced by the hottest and most energetic objects in the Universe, such as neutron stars and pulsars, supernova explosions, and regions around black holes. On Earth, gamma waves are generated by nuclear explosions, lightning, and the less dramatic activity of radioactive decay. The longest wavelengths recognized

DOI: 10.1201/9781003178477-8

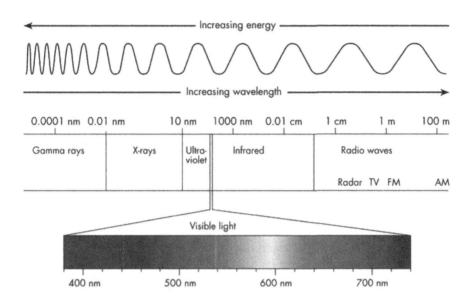

FIGURE 8.1 Electromagnetic spectrum is depicted. Wavelengths vary from thousands of kilometers down to a fraction of the size of an atomic nucleus.

in Figure 8.1 correspond to Amplitude Modulated (AM radio) carrier frequencies (see Section 7.1.9 explaining the idea of the amplitude modulation), used for broadcasting. The visible spectrum is only a tiny part of the entire electromagnetic spectrum, but even that has entrancing variety, giving colors ranging from deep purple through blue, green and yellow to deep red.

The irradiance (or intensity) of the electromagnetic wave (also called the intensity of the electromagnetic wave), denoted I equals the rate at which energy passes a unit area oriented perpendicular to the direction of propagation. The magnitude of the irradiance/intensity I of the electromagnetic wave equals to the time average of the Poynting vector (see Section 7.1.11 finalizing the previous chapter). For the plane monochromatic electromagnetic wave $\vec{E}(\vec{r},t) = \vec{E}_0 e^{j(\omega t - \vec{k}\cdot\vec{r})}$ (see Eq. (7.88a)) traveling in vacuum, it is calculated as (see Exercise (8.2)):

$$I = \frac{\varepsilon_0 c E_0^2}{2}. \tag{8.2}$$

The dimensions of the intensity of the electromagnetic wave are defined according to $[I] = \frac{W}{m^2} = \frac{J}{s \times m^2}$. When the incident light beam meets the slab of material a diversity of optical events take place, including reflection, refraction, absorption and scattering, as depicted schematically in Figure 8.2. Scattering and absorption occur in the bulk of the material. Light scattering is a term used to describe a physical process where the light beams are forced to deviate from a straight trajectory by localized non-uniformities (including particles, bubbles and radiation) in the medium through which they pass, as shown schematically in Figure 8.2. In our future treatment, we will simplify our approach and adopt, that the material is uniform; thus, the

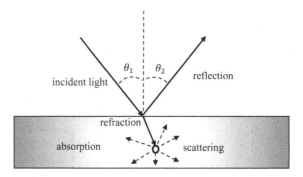

FIGURE 8.2 Optical events occurring when incident light beam encounters the slab of material. Reflection, refraction absorption and scattering are depicted schematically.

scattering is negligible. Light absorption is, in turn, a process by which light is absorbed and converted into energy (heat), and it will be addressed in detail in our future discussion.

Let us quantify the "light encounter" with the medium.[1,2] When electromagnetic wave with intensity I encounters a material, a part I_R is reflected, a part I_A is absorbed by the material, and the part I_T is transmitted. The conservation of energy yields Eq. (8.3):

$$\frac{I_R}{I_0} + \frac{I_A}{I_0} + \frac{I_T}{I_0} = 1. \tag{8.3}$$

The first term in Eq. (8.3) is called the *reflectivity* of the material, the second term is the *absorptivity* and the last is labeled the *transmittability*. Transmission of the light, in turn, includes the refraction, absorption and scattering of the light, as illustrated with Figure 8.2. All the terms appearing in Eq. (8.3) are dimensionless. Each depends on the wavelength of the electromagnetic wave, on the nature of the material and on the state of its surface, as well as its thickness (for the absorption and transmission).

Let us start from the reflection of the light by the surface. Two kinds of reflection are distinguished, namely: *specular* and *diffuse* reflection. *Specular* surfaces are microscopically smooth and flat. Light reflects from a smooth surface at the same angle as it hits the surface; thus, the condition $\theta_1 = \theta_2$ takes place, as shown in Figure 8.2. It is noteworthy that the *specular reflection law* expressed by $\theta_1 = \theta_2$ emerges from Fermat's principle to be discussed later. For a rough surface, reflected light rays scatter in various directions giving rise to diffuse reflection. This happens when the surface is rough. Most of the things we see are because light from a source has reflected off it. Now we focus on light refraction.

8.2 LIGHT REFRACTION: THE MACROSCOPIC APPROACH

When the light forming a ray moves from one medium to another – say from air to the glass slab – the incident ray changes the direction at the boundary between the media; the ray is said to undergo – refraction (see Figure 8.3). Let the index of

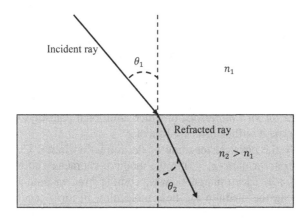

FIGURE 8.3   Refraction from a medium with index of refraction n_1 into a medium with index of refraction n_2. In this case, $n_2 > n_1$ and the refracted ray is bent toward to the normal to the boundary surface (the normal is shown with the dashed line). If n_2 had been less than n_1, the refracted ray would have bent away from the normal.

refraction of the medium with the incident ray be n_1 and that of the medium to be with the refracted ray be n_2. In optics, very roughly speaking, the refractive index (or refraction index) of an optical medium is a dimensionless number that gives the indication of the light bending ability of that medium.[3,4] The angles, that the incident and refracted rays make with the line normal to the boundary between the media are θ_1 and θ_2 (the angles are shown in Figure 8.3, the normal is depicted with the dashed line). The interrelation between the angles and the refraction indices is given by Eq. (8.4):

$$n_1 sin\theta_1 = n_2 sin\theta_2. \qquad (8.4)$$

This result found by Willibord Snell in 1621 is known as a Snell's law. Restoring historical justice demands to say that refraction and reflection of light were first studied by Ḥasan Ibn al-Haytham (965–1040) a medieval mathematician, astronomer, and physicist of the Islamic Golden Age from present-day Iraq. Ibn al-Haytham was the first to correctly explain the theory of vision, and to argue that vision occurs in the brain, pointing to observations that it is subjective and affected by personal experience.[5] He also stated the principle of least time for refraction, which would later become the Fermat's principle to be discussed below.[5]

The law of refraction may be re-written as follows:

$$\frac{sin\theta_1}{sin\theta_2} = \frac{n_2}{n_1} = n_r, \qquad (8.5)$$

where $n_r = \frac{n_2}{n_1}$ is the relative refraction index of the two media. The refraction index of air is very close to that of vacuum, which, in turn, equals to unity. Consider the situation, when the light comes from air to some transparent medium (say glass); thus, $n_1 \cong 1$ is reasonably assumed. For the usual (conventional) materials such as glass n_2

is larger than unity. Thus, $n_r = \frac{n_2}{n_1} > 1$ and consequently $\theta_1 > \theta_2$ is true; i.e., the light is bent toward the normal to the boundary surface, as depicted in Figure 8.3. Eq. (8.5) also demonstrates that when light enters a medium with a lower index of refraction, such as when a ray of light travels from water to air; the ray is bent farther away from the liner normal to the boundary. It should be emphasized that this analysis is based on the assumption that relative refraction index is given by the positive real number, which is not true for *metamaterials*, in which additional pathways of light bending become possible, as it will be discussed further.

Let us clarify the physical meaning of the refraction index. Consider the light beam coming from vacuum ($n_1 = 1$) to the medium characterized by the refraction index n_2, $n_r = \frac{n_2}{1} = n_2$. Electromagnetic wave coming from vacuum vibrates the molecules constituting the medium n_2, these molecules, in turn, generate the secondary electromagnetic wave, which recombines with the unscattered remainder of the primary wave, thus forming the transmitted electromagnetic wave.[4] The process continues over and over again as the wave advances in the transmitting medium n_2. Examine the transmitted or *refracted* beam.

The combination of the primary and secondary electromagnetic waves appears as a transmitted single net wave.[4] Assume that incident wave comes from vacuum with velocity $v_1 = c$ and that net transmitted wave propagates in the medium n_2 with the velocity v_2. Simple geometric considerations yield Eq. (8.6):

$$\frac{sin\theta_1}{v_1} = \frac{sin\theta_1}{c} = \frac{sin\theta_2}{v_2}.$$
(8.6)

Thus, we derive $\frac{sin\theta_1}{sin\theta_2} = \frac{c}{v_2}$. Involving Snell's law given by Eq. (8.4) yields Eq. (8.7):

$$n_2 = n_r = \frac{c}{v_2}; v_2 = \frac{c}{n_2}.$$
(8.7)

For the usual, natural materials $n_2 > 1$, thus the propagation of the refracted wave in the medium n_2 is macroscopically seen as "slowing" of the light to the velocity $v_2 = \frac{c}{n_2}$. And it should be emphasized that we speak about the *phase velocity* of the light. Actually, this slowing arises as result of combination of original/incident and secondary electromagnetic wave, and the effect of slowing should be taken with much care, *cum grano salis*; actually, light always propagates with the velocity of c. This is strongly emphasized in the Feynman Lectures on Physics, from which we are quoting: "for a piece of glass, you might think: You should say it (propagating beam) is retarded at the speed c/n. That, however, is not right, and we have to understand why it is not. It is approximately true that light or any electrical wave *does appear* to travel at the speed c/n through a material whose index of refraction is n, but the fields are still produced by the motions of *all* the charges – including the charges moving in the material – and with these basic contributions of the field travelling at the ultimate velocity c. Our problem is to understand how the *apparently* slower velocity comes about".[6] We'll address this problem in detail in this chapter, when the *microscopic origin* of the refraction will be discussed. Now we adopt the following macroscopic approach: *it looks like* that the refracted beam travels in the medium n_2 with the reduced velocity $v_2 = \frac{c}{n_2}$.

What are the changes which take place in the refracted electromagnetic wave in addition to the change in its direction and "slowing"? The wavelength of the refracted electromagnetic wave decreases because the frequency v is unchanged while the speed is decreased.[4] Consider the electromagnetic wave, which comes from vacuum with the wavelength $\lambda_0 = \frac{c}{v}$ to the medium n_2. The wavelength λ in the medium n_2 is calculated as follows:

$$\lambda = \frac{v_2}{v} = \frac{c}{n_2 v} = \frac{\lambda_0}{n_2}. \tag{8.8}$$

Thus, we conclude that the wavelength of the refracted electromagnetic wave λ is decreased when compared that coming from vacuum λ_0 according to $\lambda = \frac{\lambda_0}{n_2}$, $n_2 > 1$.[4]

8.3 ENERGY IN REFLECTION AND REFRACTION

Refraction is generally accompanied by reflection, as it follows from Eq. (8.3). At the boundary between media, this energy is apportioned among the reflected and refracted rays.[3] The reflected and refracted energies are taken right at each side of an interface/boundary and do not account for attenuation of a wave in an absorbing medium following transmission or reflection; thus, the energy dissipation due to absorption is neglected.[3] Distribution of energy between the reflected and refracted beams depends on the incident angle θ_1 (see Figures 8.2 and 8.3). Consider the simplest, particular case, when light is perpendicularly incident ($\theta_1 = \frac{\pi}{2}$) on a surface that separates a medium of index of refraction n_1 from a medium of refraction n_2. The solution of Maxwell's equations (Eqs. (7.82a–d)), considering the boundary conditions, leads to the conclusion, that the intensity of the reflected light I_R is related to the incident intensity I_0 (see Eq. (8.3), and consider $I_A = 0$) is given by Eq. (8.9) (see Reference 4):

$$\frac{I_R}{I_0} = \frac{(n_2 - n_1)^2}{(n_2 + n_1)^2}. \tag{8.9}$$

For light perpendicularly incident from air ($n_1 = 1.0$) into glass ($n_2 = 1.5$) only 4% of the incident light is reflected. Generally, the intensity of the reflected light varies with the angle of incidence $\frac{I_R}{I_0} = \frac{I_R}{I_0}(\theta_1)$ and the ratio $\frac{I_R}{I_0}(\theta_1)$ is given by the Fresnel formulas.[4]

8.4 TOTAL INTERNAL REFLECTION

For some incident angles all the incident energy is contained in the reflected ray (see Section 8.3). This situation known as a total internal reflection can occur only when light travels from a medium with a larger index of refraction toward a medium with a smaller index of refraction, as illustrated with Figure 8.4, such as when light passes from water or glass toward air. Simple geometry explains this phenomenon.

Consider a light beam incident from a medium with an index of refraction n_1 to a medium with an index of refraction n_2 and the condition $n_1 > n_2$ takes place. Snell's law expressed with Eq. (8.5) may be re-shaped as follows: $\sin\theta_2 = \frac{n_1}{n_2}\sin\theta_1$. Because

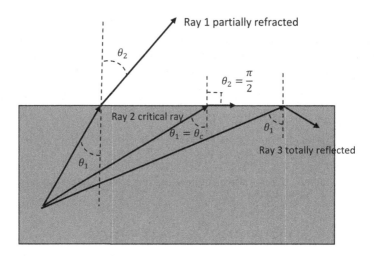

FIGURE 8.4   Total internal reflection is illustrated.

the factor $\frac{n_1}{n_2}$ is larger than unity ($n_1 > n_2$ is adopted), angle θ_2 reaches $\frac{\pi}{2}$ before θ_1 does as θ_1 increases. Figure 8.4 illustrates what happens for various values of the incident angle θ_1. When $\theta_2 = \frac{\pi}{2}$, the ray in medium 2 skims along the boundary/interface separating the media (see Figure 8.4). This occurs when the incident angle θ_1 reaches a critical value, denoted θ_c, such that $\frac{n_1}{n_2} sin\theta_c = sin\frac{\pi}{2} = 1$. This leads to Eq. (8.10):

$$sin\theta_c = \frac{n_2}{n_1}. \tag{8.10}$$

When θ_1 exceeds θ_c, there is no angle θ_2 that can satisfy Snell's law. When $n_2 = 1$ (corresponding to air), we obtain for the critical angle: $sin\theta_c = \frac{1}{n_1}$. The electromagnetic energy carried by the incident ray must go somewhere, and the ray is reflected. There is no diminution of the intensity of the intensity of the reflected ray (again the absorption is completely neglected): the reflection is total.[3]

The most important technological field exploiting the effect of the total internal reflection is the communication with the optical fibers (and this is the very essential field of the materials engineering science).[7] The physical principle enabling this kind of communication is straightforward: a transparent glass or polymer fiber will serve as a conductor of light if any ray inside the fiber undergoes total internal reflection upon striking the side of the fiber, as illustrated with Figure 8.5. Figure 8.5 depicts a ray coming from air ($n = 1$) to the fiber with the refraction index n_f. The ray comes to the fiber with an incident angle θ_i (see Figure 8.5); the interrelation between the angles according to Snell's law (Eq. (8.4)) is $sin\theta_f = \frac{sin\theta_i}{n_f}$. This ray will strike the wall of the fiber at the angle $\frac{\pi}{2} - \theta_f$ with the normal to the wall (see Figure 8.5). We have the total internal reflection if $n_f sin\left(\frac{\pi}{2} - \theta_f\right) > 1$, in other words, when $n_f cos\theta_f > 1$. We derive:

$$n_f cos\theta_f = n_f \sqrt{1 - sin^2\theta_f} = n_f \sqrt{1 - \frac{sin^2\theta_i}{n_f^2}} = \sqrt{n_f^2 - sin^2\theta_i} > 1. \tag{8.11}$$

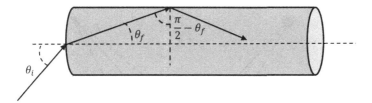

FIGURE 8.5   Total internal reflection occurring within the optical fiber is shown. The light enters the fiber at angle θ_i.

Considering $sin^2\theta_i \leq 1$ yields Eq. (8.12):

$$\sqrt{n_f^2 - sin^2\theta_i} \geq \sqrt{n_f^2 - 1}.$$ (8.12)

Thus, the condition of the total internal reflection is guaranteed with Eq. (8.13):

$$\sqrt{n_f^2 - 1} > 1.$$ (8.13)

The largest value of $sin\theta_i$ is one; thus, Eq. (8.13) is a condition for total internal reflection for all the beams that enter the fiber. Eq. (8.13) is satisfied for any material with $n_f > \sqrt{2}$. A typical optical fiber has an index of refraction of $n \cong 1.62$ which is larger than the critical value $\sqrt{2}$. It should be emphasized that once a ray is in the fiber, it remains inside even if the fiber curves.[3,7]

Actually, the structure of optical fibers is more complicated. An optical fiber consists of three concentric elements, the core, the cladding and the outer coating, often called the buffer.[7] The core is usually made of glass or plastic. The core is the light-carrying portion of the fiber. The cladding surrounds the core. The cladding is made of a material with a slightly lower index of refraction than the core.[7] This difference in the indices causes total internal reflection to occur at the core-cladding boundary along the length of the fiber; thus, light is transmitted down the fiber and does not escape through the sides of the fiber.[7]

8.5   FERMAT'S PRINCIPLE

The laws of reflection and refraction, and the manner in which light propagates in the material can be seen within an entirely different and intriguing paradigm, constituted by Fermat's principle. Fermat Principle (in its initial formulation, which was essentially corrected recently, as it will be discussed below) states that, the actual path between two points taken by a beam of light is the one as traversed in the least time.[4] This principle is one of the variational principles of physics, which play fundamental, basic role in its structure. An excellent introduction to the variational principles of physics you find in Chapter 19 of the second volume of the Feynman Lectures on Physics.[6] The first (and incorrect!) version of the Fermat Principle was suggested by Hero from Alexandria, who lived in Egypt during the Roman era. Hero suggested that *the path taken by the light in going from some point S to a point P via*

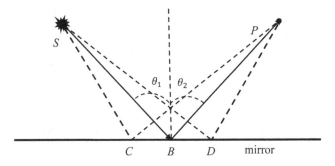

FIGURE 8.6 Testing various pathways possible under reflection of the light beam, connecting the source located in point *S* and point *P*. The pathway *SBP* is the shortest one.

reflecting surface is the shortest possible one (see Figure 8.6). Let us test different pathways, which are possible under reflection of the beam generated by the point source *S*, reflected from the mirror and coming to the point *P*, as shown in Figure 8.6. These pathways are labeled *SCP*, *SBP* and *SDP* in Figure 8.6. Presumably, only one of these pathways will have a physical reality. The pathway *SBP* is the shortest possible one, corresponding to $\theta_1 = \theta_2$; indeed, it is the shortest from possible pathways (prove it), as it was seen by Hero from Alexandria. Hero's discovery stood alone, until Ḥasan Ibn al-Haytham already mentioned in Section 8.2 rediscovered this idea. Fermat in 1657 suggested the much more comprehensive *Principle of Least Time*. Pierre de Fermat (1607–1665), a lawyer at the Parlement of Toulouse, France, was a French mathematician who is given credit for early developments that led to infinitesimal calculus, analytic geometry, probability and optics. He is best known for his Fermat's Last Theorem in number theory, which he described in a note at the margin of a copy of Diophantus' Arithmetica.[8] The *Principle of Least Time* suggested by Fermat encompassed both reflection and refraction. Fermat suggested that a light beam actually does not take a minimal spatial path between the points, as it was suggested by Hero from Alexandria. According to Fermat, the actual path between two points taken by a beam of light is the one that is traversed in the *least time*.[4] Today, we know, that not only least but also maximal time span of the light propagation should be considered, as it will be discussed below.

It should also be mentioned that the Fermat Principle may be re-shaped in terms of the *optical path length* (usually abbreviated OPL), which has to be clearly distinguished from the path length. Optical path length is the length that light needs to travel through a vacuum to create the same phase difference as it would have when traveled through a given medium. It is calculated as $\int_C n\,ds$, where *n* is the local refractive index as a function of distance along the path *C*. The Fermat Principle in the terms of optical path length is re-formulated as follows: the OPL of the actual ray is either an extremum (a minimum or a maximum) with respect to the OPL of adjacent paths or equal to the OPL of adjacent paths.

Now let us see how Snell's law arises from the principle of the least time (Fermat's Principle). We have to minimize τ, which is the transit time from source (labeled in Figure 8.7 with *S*) to point *P*. We minimize τ with respect to variable *x* (see

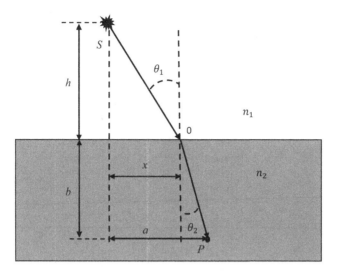

FIGURE 8.7 Fermat's principle applied to refraction. The transit time τ from source (labeled in with S) to point P should be minimized.

Figure 8.7). In other words, changing x shifts 0, thus, varying the ray from S to P, and testing various possible pathways of the light propagation (see Figure 8.7). The smallest transit time will then presumably coincide with the actual path.[4] The total propagation time from S to P is given by Eq. (8.14):

$$\tau = \frac{SO}{v_1} + \frac{OP}{v_2},\tag{8.14}$$

where v_1, v_2 are the velocities of the light propagation in the media n_1, n_2 correspondingly. Geometrical considerations (see Figure 8.7) lead to:

$$\tau(x) = \frac{\sqrt{h^2 + x^2}}{v_1} + \frac{\sqrt{b^2 + (a-x)^2}}{v_2}.\tag{8.15}$$

To minimize $\tau(x)$ with respect to variations in x, we set $\frac{d\tau(x)}{dx} = 0$; this, in turn, yields:

$$\frac{d\tau(x)}{dx} = \frac{x}{v_1\sqrt{h^2 + x^2}} + \frac{-(a-x)}{v_2\sqrt{b^2 + (a-x)^2}} = 0.\tag{8.16}$$

Simple trigonometric considerations yield:

$$\frac{sin\theta_1}{v_1} = \frac{sin\theta_2}{v_2},\tag{8.17}$$

which is an exact expression of Snell's law (compare with Eq. (8.6), please).

We already mentioned that the Fermat Principle, as it was suggested by Pierre de Fermat, asserts that the actual path between two points taken by a beam of light is the

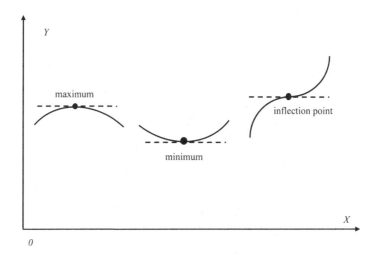

FIGURE 8.8 Three kinds of stationary points are depicted: maximum, minimum and the inflection point.

one that is traversed in the least time, is not exact. Fermat's principle in its modern form states that a light ray in going from point S to point P must traverse an optical path length that is *stationary* with respect to variations of that path.[4] Stationary path is the path taken by a light between two points, which *minimizes* or *maximizes* the time taken by the light ray. Thus, pathways which *maximize* the time taken by the light beam in going from point S to point P are possible. Formulation of the Fermat Principle in terms of the minimum of wave travel time between two points is not correct in general. The correct formulation involves the extremum of the total optical length, with the optical length for the wave propagation through left-handed materials taken to be negative. The rigorous formulation of the Fermat Principle sounds as follows: actual pathway of the light propagation corresponds to the local extremum of the optical path length, as it was worded by Victor Vassalage, a pioneer in the field of metamaterials.[9]

Or rigorously mathematically speaking: a light ray in going from point S to point P must traverse an optical path length that is stationary with respect to variations of this path. By a stationary value of the function $f(x)$ we mean one for which the slope of $f(x)$ vs. x is zero or equivalently where the function $f(x)$ has a maximum, minimum or a point of inflection with a horizontal tangent as depicted in Figure 8.8. The kind of extremum (maximum, minimum or inflection point) depends on the actual values of the refraction index of the medium.

8.6 MICROSCOPIC ORIGIN OF THE REFRACTION INDEX: CALCULATION OF THE ATOMIC POLARIZABILITY

Until now, we treated the refraction index within the pure macroscopic paradigm. In this section, we will try to understand the microscopic origin of the refraction index. We also will explain the effect of "slowing" of the light quantified by the refraction index (see Eq. (8.7)). In our treatment we keep the approach developed by Richard Feynman in Chapter 31 of his Feynman Lectures of Physics, which are strongly recommended to the reader of our book (see Reference 6). This approach is based on the following fundamental assumptions:

i. The total electric field in any physical circumstance can always be represented by the sum of the fields from all the charges in the universe. This principle is also known as the principle of superposition.

ii. The field generated by a single charge is given by its acceleration evaluated with a retardation at the speed c, always (for the radiation field).

For a piece of polymer, glass or other dielectric material you might think: "This is not true, you should modify all this. You should say it is retarded at the speed $\frac{c}{n}$". That, however, is not right, and we have to understand why it is not.[6] It is approximately true that light or any electrical wave does appear to travel at the speed $\frac{c}{n}$ through a material whose index of refraction is n, but the fields are still produced by the motions of all the electrical charges, including the charges moving in the material – and with these basic contributions of the field traveling at the ultimate and velocity c, which is a velocity of light in vacuum. Our problem is to understand how the apparently slower velocity comes about.[6] Consider the situation, illustrated in Figure 8.9. The source of electromagnetic waves (say lightbulb) is located in point S. The electromagnetic wave (light) generated by the bulb passes through the slab of the dielectric material (glass or polymer), as shown in Figure 8.9. The question is: what electric field in point P (see Figure 8.9) is. According to the first of the aforementioned principles, the total electric field in point P $\vec{E}$ is given by Eq. (8.18):

$$\vec{E} = \sum_{all\,charges} \vec{E}_{each\,charge},\tag{8.18}$$

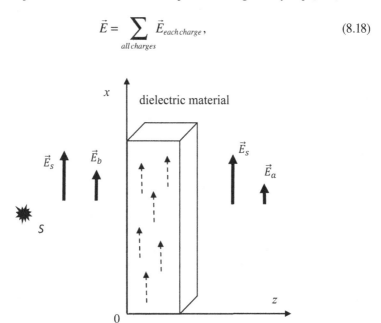

FIGURE 8.9   Microscopic origin of the refraction index n. $\vec{E}_s$, $\vec{E}_b$ and $\vec{E}_a$ are the electrical field produced by the source, the reflected field and field produced by the charge located in the dielectric correspondingly. Dashed arrows depict the dipoles constituting the dielectric material.

which may be re-written as follows:

$$\vec{E} = \vec{E}_S + \sum_{all\,other\,charges} \vec{E}_{each\,charge}, \qquad (8.19)$$

where $\vec{E}_S$ is the field due to the source/lightbulb alone and would be precisely the field at point P if there were no material present. We expect the field at P to be different from $\vec{E}_S$ if there are any other moving charges.[6] What is the origin of these charges? Why should there be charges moving in the glass or another dielectric material? All material consists of atoms or molecules, which contain electrons. When the electric field *of the source* acts on these atoms it drives the electrons up and down, because it exerts a force on the electrons. And moving electrons generate a field – they constitute new radiators/ sources of the electromagnetic field. These new radiators are related to the source S, because they are driven by the field of the source.[6] The total field is not just the field of the source S, but it is modified by the additional contribution from the other moving charges (see Figure 8.9). This means that the field is not the same as the one which was there before the glass was there, but is modified, and it turns out that it is modified in such a way that the field inside the glass appears to be moving at a different speed.[4,6]

There is another effect caused by the motion of the charges in the dielectric plate. These charges will also radiate waves back toward the source S (see Figure 8.9). This backward-going field is the light we see reflected from the surfaces of transparent materials. It does not come from just the surface. The backward radiation comes from everywhere in the interior, but it turns out that the total effect is equivalent to a reflection from the surfaces. These reflection effects are beyond our approximation at the moment because we restrict ourselves to a calculation for a material with the refraction index so close to unity that very little light is reflected.

Accurate problem of the calculation of the resulting field is extremely challeng- ing. If we consider a particular charge with a slab of dielectric material, depicted in Figure 8.9, it feels not only the source S, but it feels *all* of the charges that are mov- ing. It feels, in particular, the charges that are moving somewhere else in the glass. So, the total field which is acting on a *particular charge* is a combination of the fields from the other charges, *whose motions depend on what this particular charge is doing,* and the calculation of this field is an extremely complicated task.

Let us start from a much simpler situation, occurring when the effects from the other atoms are very small relative to the effects from the source. In other words, we take a material in which the total field is not modified very much by the motion of the other charges. That corresponds to a material in which the index of refraction is very close to unity, $n \cong 1$. This is true, if the density of the atoms is very low. The simplest material exemplifying the rare material is gas.

Consider gas comprising $\tilde{n}$ particles per unit volume (we use $\tilde{n}$ as a symbol for the number density of particles to be distinguished from the refractive index n). Now we make the main and extremely instructive assumption; namely, we adopt that each particle behaves as a harmonic oscillator (see Sections 7.1–7.1.5 of the previ- ous chapter). Now it is becoming clear, why we spent a great effort in the analysis of the motion of the harmonic oscillator. We use a model of an atom or molecule in which the electron is bound with a force proportional to its displacement (as though the electron were held in place by an effective Hookean spring, see Section 7.1.1). Of course, this is a very simple model, however, the correct quantum mechanical theory gives results equivalent to this model (in simple cases). So, our simple model looks like that: electromagnetic wave $E = E_0 e^{j\omega t}$ produced by the source encounters electron and makes it to oscillate. Thus, the force acting on the electron is given by $F = eE = eE_0 e^{j\omega t}$. This force excites the forced oscillations of the electron, addressed in Section 7.1.6 of the previous chapter.

We also consider the damping force acting on the electron. Such a force corresponds to a resistance to the motion, that is, to a force proportional to the velocity of the electron $f_{fr} = -b\dot{x} = -b\frac{dx}{dt}$ (see Section 7.1.5). Then, the equation of motion of the electron is (see Section 7.1.6):

$$-kx - b\dot{x} + eE_0 e^{j\omega t} = m_e \ddot{x}, \qquad (8.20)$$

where m_e is the mass of the electron. Eq. (8.20) may be re-written, as follows:

$$\ddot{x} + 2\beta\dot{x} + \omega_0^2 x = \frac{eE_0}{m_e} e^{j\omega t}, \qquad (8.21)$$

where $\omega_0 = \sqrt{\frac{k}{m_e}}$, $\beta = \frac{b}{2m_e}$. The solution of this equation is given by Eq. (7.36) supplied in the previous chapter, where $f_0 = \frac{eE_0}{m_e}$. It is also usual to re-label $2\beta = \gamma$ in the theory of forced electromagnetic oscillations (γ is called the dissipation constant, see Reference 6). As it was already mentioned in Chapter 7, the solution of Eq. (8.21) supplied by Eq. (7.36) is built from two terms: the first term of the solution, given by Eq. (7.36) represent exponential effects due to the friction, and it is important only at the first stage of oscillations. These effects emerging from friction will die with time (the characteristic time of decay of the effects due to friction $\tau \sim \frac{1}{\beta} \sim \frac{2}{\gamma}$). The second term represents harmonic motion with the frequency of the driving force ω, namely $x = x_0 e^{j\omega t}$. Substituting $x = x_0 e^{j\omega t}$ into Eq. (8.21), considering $2\beta = \gamma$, $E = E_0 e^{j\omega t}$ and exploiting the mathematical technique of complex numbers developed in Section 7.1.2, we obtain for the steady-state solution of Eq. (8.21) the following expression:

$$x(t) = \frac{e/m_e}{-\omega^2 + j\gamma\omega + \omega_0^2} E_0 e^{j\omega t} = \frac{e/m_e}{-\omega^2 + j\gamma\omega + \omega_0^2} E. \qquad (8.22)$$

The dipole moment induced by electromagnetic field dipole equals $p = ex$; thus, Eq. (8.22) yields for the vector of the induced dipole moment:

$$\vec{p} = \frac{e^2/m_e}{-\omega^2 + j\gamma\omega + \omega_0^2} \vec{E}. \qquad (8.23)$$

Electrodynamics of continuous media supplied the relation between the induced dipole moment and the electrical field resumed by Eq. (8.24) (for the precise definition of the dipole moment, see Section 6.2.2 and Figure 6.3B):

$$\vec{p} = \varepsilon_0 \alpha(\omega) \vec{E}, \qquad (8.24)$$

where $\alpha(\omega)$ is called the atomic polarizability (see Section 6.2.2; Eq. (8.24) follows the definition of atomic polarizability accepted in Reference 6, where $\alpha(\omega)\vec{E}$ is multiplied by ε_0). Comparing Eqs. (8.23) and (8.24), we calculate:

$$\alpha(\omega) = \frac{e^2/\varepsilon_0 m_e}{-\omega^2 + j\gamma\omega + \omega_0^2}. \tag{8.25}$$

Now let us calculate the refractive index for the simplest medium, i.e., an ideal gas. First of all, we have to consider the interrelation between the refraction index n and the atomic polarizability $\alpha(\omega)$ of the dielectric material, which is expressed with Eq. (8.26):

$$n^2 = 1 + \tilde{n}\alpha(\omega). \tag{8.26}$$

Thus, we calculate the refractive index of a gas:

$$n^2 = 1 + \frac{\tilde{n}e^2}{m_e\varepsilon_0} \frac{1}{-\omega^2 + j\gamma\omega + \omega_0^2}. \tag{8.27}$$

Actually, the situation is more complicated even for gases. Following Reference 6, we state that the quantum mechanical solution for the motions of electrons in atoms gives a similar answer except with the following modifications. The atoms have several natural frequencies, each frequency with its own dissipation constant γ. Also, the effective "strength" of each mode is different, which we can represent by multiplying the polarizability for each frequency by a dimensionless strength factor f, which is a number we expect to be of the order of unity. Representing the three parameters ω_0, γ, and f by ω_{0k}, γ_k and f_k for each mode of oscillation, and summing over the various modes, we modify Eq. (8.27) and re-shape it as follows:

$$\alpha(\omega) = \frac{e^2}{\varepsilon_0 m_e} \sum_k \frac{f_k}{-\omega^2 + j\gamma_k\omega + \omega_{0k}^2}. \tag{8.28}$$

Eq. (8.28) is true for a single atom. What do we have for gas built of a number of atoms? If $\tilde{n}$ is the number of atoms per unit volume in the material, the polarization P of the gas (which is the volumetric density of permanent or induced electric dipole moments in a dielectric material) is supplied by Eq. (8.29):

$$\vec{P}(\omega) = \tilde{n}\vec{p} = \varepsilon_0\tilde{n}\alpha(\omega)\vec{E}, \tag{8.29}$$

where $\alpha(\omega)$ is given by Eq. (8.28). Let us inspect closely Eqs. (8.28) and (8.29). We recognize that (i) the atomic polarizability α and volumetric polarizability P are expressed by complex numbers; (ii) both parameters are frequency dependent. These conclusions are of crucial importance for the development of metamaterials, as it will be demonstrated in the next chapter.

In the dense dielectric material there exist additional complications and challenges: the neighboring atoms are so close, there are strong interactions between them. The internal modes of oscillation are, therefore, modified.[6] The natural frequencies of the atomic oscillations are spread out by the interactions, and they are usually quite heavily damped; the resistance coefficient becomes quite large. So, the values of ω_{0k} and γ_k of the solid will be quite different from those of the free atoms.[6]

There is one more additional and essential complication in dense dielectric materials. The electric field acting on atoms is different from the external electric field E, because in dense materials there is also the field produced by other atoms in the vicinity, which may be comparable to the external field, which should be carefully considered (see Reference 6). With these reservations, we can still represent α, at least approximately, by Eq. (8.28). However, the principles underlying derivation of the refraction index of the dense dielectric materials remain the same, as discussed in Reference 6. The rigorous calculation of the refractive index of the dense dielectric materials yields for the refractive index n (see Eqs. (8.28), (8.29) and Reference 6):

$$3\frac{n^2-1}{n^2+2} = \frac{\tilde{n}e^2}{\varepsilon_0 m_e}\sum_k \frac{f_k}{-\omega^2 + j\gamma_k\omega + \omega_{0k}^2}. \tag{8.30}$$

Thus, we came to two extremely important conclusions, arising from Eq. (8.30): (i) the refractive index of a dense dielectric material is supplied by complex numbers; (ii) the refractive index of a dense dielectric material is frequency dependent.

8.7 PHYSICAL MEANING OF THE REAL AND IMAGINARY PARTS OF THE REFRACTION INDEX

In the previous section, we came to the conclusion that the refraction index of a dielectric material n is expressed with the complex number (see Eq. (8.30)). We demonstrated in Section 7.1.2 that the complex number z may be presented as a sum of the real and imaginary parts: $z = z_r + jz_i = Re(z) + jIm(z)$. So let us re-write the complex refraction index as follows:

$$n = n_R - jn_I. \tag{8.31}$$

We write $n = n_R - jn_I$ with a minus sign, as suggested in Reference 6, so that n_I will be a positive quantity in all ordinary optical materials. In ordinary inactive materials, that are not, like lasers, light sources themselves, the dissipative constant γ is a positive number, and that makes the imaginary part of n negative. Consider the plane electromagnetic wave propagating in direction z, $E_x = E_0 e^{j(\omega t - kz)}$ in the medium characterized by the refraction index n (the orientation of axes is shown in Figure 8.9). We use the following formulas: $v = \frac{\omega}{k}$ (see Eq. (7.52)) and $v = \frac{c}{n}$ (see Eq. (8.7)) and perform the following simple transformations:

$$E_x = E_0 e^{j(\omega t - kz)} = E_0 e^{j\omega\left(t-\frac{k}{\omega}z\right)} = E_0 e^{j\omega\left(t-\frac{z}{v}\right)} = E_0 e^{j\omega\left(t-\frac{nz}{c}\right)}. \tag{8.32}$$

Now we substitute $n = n_R - jn_I$ into Eq. (8.32) and derive:

$$E_x = E_0 e^{-\omega\frac{n_I z}{c}} e^{j\omega\left(t-\frac{n_R z}{c}\right)}. \tag{8.33}$$

The term $e^{j\omega(t-\frac{n_R z}{c})}$ represents a wave traveling with the speed $\frac{c}{n_R}$, so n_R represents what we normally think of as the index of refraction. However, the *amplitude* of this wave $\tilde{E}$ is given by Eq. (8.34):

$$\tilde{E} = E_0 e^{-\omega\frac{n_I z}{c}}. \tag{8.34}$$

which decreases exponentially with z. Thus, the imaginary part of the index represents the attenuation of the wave due to the energy losses in the atomic oscillators. The intensity of the wave is proportional to the square of the amplitude (see Eq. (8.2)), so the intensity of the transmitted electromagnetic wave scales as:

$$I \sim e^{-2\omega\frac{n_I z}{c}}, \tag{8.35}$$

which is often presented as follows:

$$I \sim e^{-\kappa z}, \tag{8.36}$$

where $\kappa = 2\omega\frac{n_I}{c}$ is known in optics as the absorption coefficient. Eq. (8.36) is known in optics as the Bouguer–Beer–Lambert law of the light absorption,[4] and it written eventually as follows:

$$\frac{I}{I_0} = e^{-\kappa z}, \tag{8.37}$$

where $\tilde{I}_0$ is the intensity of the light at the beginning of the travel.

We came to a very important conclusions: (i) the refractive index $n = n_R - jn_I$ is given by a complex number; (ii) the real part of the refraction index n_R is responsible for the propagation of the electromagnetic wave; whereas the imaginary part of the refractive index n_I is responsible for the absorption of the electromagnetic wave (see Eq. (8.33)).

8.8   THE DRUDE MODEL

Now we address microscopic origin of refraction index of metals. There exist two general approaches to the microscopic structure of metals, namely the classical and the quantum mechanics-based ones. Reference 10 is strongly recommended for both approaches. Of course, the quantum mechanics-based approach is more rigorous and adequate; however, our in our treatment we restrict ourselves with the classical approach, which is known in the solid state physics, as the Drude-Lorentz theory (or for a sake of brevity, the Drude model). This approach, being erroneous, remains, however, very simple and understandable, and results in the qualitatively reasonable results. The core assumptions made in the Drude model are the following:

- The Drude model considers the metal is formed of a collection of positively charged ions from which a number of "free electrons" were detached. These may be thought to be the valence electrons of the atoms that have become delocalized due to the electric field of the other atoms.

- Drude adopted that electrons within a metal behave as an ideal gas.
- The model assumes that free electrons move in parallel in the random/ thermal and ordered/drift motions. Thermal motion is relatively rapid, whereas the drift motion is relatively slow.
- The Drude model neglects long-range interaction between the electron and the ions or between the electrons; this is called the independent electron approximation.
- The only interaction of a free electron with its environment was treated as being collisions with the impenetrable ion cores.
- The electrons move in straight lines between one collision and another; this is called free electron approximation.
- After a collision event, the distribution of the velocity and direction of an electron is determined by only the local temperature and is independent of the velocity of the electron before the collision event. The electron is considered to be immediately at equilibrium with the local temperature after a collision.

Now let us apply the electrical field E to the metal. Force acting on a free electron is eE; thus, the Newton Second Law for the electron is given by Eq. (8.38):

$$eE = m_e a, \qquad (8.38)$$

where a is the acceleration of electron; its value may be estimated as $a = \frac{v_{max}}{\tau}$, where v_{max} is the maximal velocity of the accelerated electron and τ is the average time between two consequent collisions (the meaning and value of τ will be discussed below). Substitution of this estimation into Eq. (8.38) yields for the maximal velocity of the electron:

$$v_{max} = \frac{eE\tau}{m_e} \qquad (8.39)$$

We adopt that electron moves with the constant acceleration; thus, the average velocity of its drift motion, denoted $\langle v \rangle$ is given by $\langle v \rangle = \frac{v_{max}}{2}$. Considering Eq. (8.39) yields for the average velocity of electron:

$$\langle v \rangle = \frac{eE\tau}{2m_e}. \qquad (8.40)$$

This averaged velocity labeled $\langle v \rangle$ is called in the solid state physics "the drift velocity". The drift velocity is the average velocity attained by charged particles, such as electrons, in a material due to an electric field. The value of the drift velocity of electrons is estimated in Eq. (8.20), and it is demonstrated that that the drift velocity of electrons is much smaller than the velocity of its thermal motion; this interrelation between velocities is an important feature of the completely classical Drude theory/model.

According to a quantum theory, an electron in a conductor will propagate randomly at the Fermi velocity, resulting in an average velocity of zero. Applying an electric field adds to this random motion a small net flow in one direction; this is a *drift*. The drift velocity is the velocity of the motion of electrons ordered by an external electric field E. Just this drift velocity is responsible for the electric current. The density of the current J is calculated as follows:

$$J = \frac{I}{S} = e\langle v\rangle \tilde{n}, \tag{8.41}$$

where $I = \frac{dQ}{dt}$ is the electrical current, defined as the net rate of flow of electric charge Q through a surface S; the unit of electric current in the International System of Units (SI) is the ampere, denoted A, $[I] = A$, in other words one ampere is equal to 1 coulomb moving past a point in 1 second, ampere is the base unit in the SI system; $\tilde{n}$ is the charge-carrier number density $\tilde{n} = \frac{N}{V}$, where N is the total number of charge carriers/ electrons in a volume V, $[\tilde{n}] = \mathrm{m}^{-3}$. Substitution of Eq. (8.40) into Eq. (8.41) results in:

$$J = \frac{\tilde{n}e^2\tau}{2m_e} E. \tag{8.42}$$

Compare Eq. (8.42) with the well-known Ohm Law, given by Eq. (8.43):

$$J = \sigma E, \tag{8.43}$$

where σ is the specific conductivity of the metal. Recall that the conductivity is reciprocal to resistivity $\sigma = \frac{1}{\rho}$, $\sigma = \Omega^{-1}\mathrm{m}^{-1}$; $[\rho] = \Omega\mathrm{m}$ in SI. It is immediately recognized that Eq. (8.42) represents the Ohm law. Thus, we conclude that the Drude theory/model yields the Ohm law. Comparing Eq. (8.42) to Eq. (8.43) yields for the specific conductivity of the metal σ:

$$\sigma = \frac{\tilde{n}e^2\tau}{2m_e}. \tag{8.44}$$

What is the physical meaning of the τ, appearing in Eq. (8.44) and defined as the average time between two consequent collisions? It may be very roughly estimated as:

$$\tau \cong \frac{\lambda}{v_{th}}, \tag{8.45}$$

where λ is the average distance between two consequent collisions (also called mean free path) and v_{th} is the average velocity of the thermal motion of electrons. Why do just v_{th} appear in Eq. (8.45)? The fact is that the thermal velocity is much larger than the drift velocity of electrons (see Exercises (8.17) and (8.20), which are very important for understanding the Drude model). That the averaged velocity of the motion of electrons is very close to that of the thermal motion. Substitution of Eq. (8.45) into Eq. (8.44) yields for the specific conductivity of metals σ:

$$\sigma = \frac{\tilde{n}e^2\lambda}{2m_e v_{th}}. \tag{8.46}$$

Eqs. (8.44) and (8.46) represent the remarkable achievements of the Drude model:

i. Eqs. (8.44) and (8.46) express the macroscopic characteristic of metals, namely, their specific conductivity σ via their microscopic parameters: $e, m_e, \tau, \lambda, v_{th}$.

ii. The quantities e and m_e are independent on the type of metal. The parameters appearing in Eqs. (8.44) and (8.45) are independent on the electrical field E, hence, the specific conductivity of metals as the fixed temperature is constant; thus, supplying the microscopic foundation to the Ohm law (see Eq. (8.42)).

The Drude model possesses an impressive list of scientific achievements, namely:

i. It explains the heating of conductors by electrical currents governed by the Joule-Lenz law.

ii. It explains the Wiedemann–Franz law states that the ratio of the electronic contribution of the thermal conductivity to the electrical conductivity of a metal is proportional to the temperature of the conductor.[10]

iii. It explains qualitatively the Hall effect, which is s the production of a potential difference (the *Hall voltage*) across an electrical conductor that is transverse to an electric current in the conductor and to an applied magnetic field perpendicular to the current.[10]

As any other physical model, the Drude model has shortcomings, some of which are listed below:

i. The Drude model assumes that the electrons move independently of each other, ignoring the effects of electron-electron interactions. In reality, the electrons in a metal interact with each other through Coulomb forces, which can lead to deviations from the predictions of the Drude model.

ii. The Drude model assumes that the scattering of electrons is purely elastic, meaning that the energy of the scattered electrons is the same as the energy of the incident electrons. In reality, electron scattering can be inelastic, leading to the emission or absorption of energy and a change in the energy of the electrons.

iii. The Drude model assumes that the electron mass is constant and independent of the electron's energy. In reality, the effective mass of the electrons can depend on their energy and on the crystal structure of the metal.

iv. The Drude model assumes that the electrons are scattered by the stationary ions in the metal lattice. However, in reality, the ions are not completely stationary but are subject to thermal vibrations, which can lead to additional scattering of the electrons.

v. And the most important thing: Drude model does not take into account the effects of quantum mechanics, which can become important at low temperatures and small length scales.

Anyway, the Drude theory is a remarkable physical model, explaining a diversity of physical phenomena, and we will use Eqs. (8.44) and (8.46) for understanding the properties of the metamaterials.

8.9 THE PLASMA FREQUENCY

Now we have to develop one more notion, necessary for understanding the microscopic origin of the refraction index in metals, and that is the plasma frequency. We already met this notion when we discussed propagation of electromagnetic waves in ionosphere (see Section 7.1.9). Consider an electrically neutral plasma in equilibrium, consisting of a set of positively charged ions and negatively charged electrons. If one displaces by a tiny amount an electron or a group of electrons with respect to the ions at rest, the Coulomb force pulls the electrons back, acting as a restoring force, giving rise to harmonic oscillations. Let us supply the quantitative analysis to this effect. Consider the rectangular metallic prism filled with the ions and a free electron gas with the volume density $\tilde{n}$; the cross-section of the prism is S, as depicted in Figure 8.10. We displace the free electron gas to the right at a distance δx, as shown in Figure 8.10. Thus, an excess, positive, electrical charge with a surface density $+\tilde{\sigma}$ is exposed at the right side of the prism; whereas the negative electrical charge $-\tilde{\sigma}$ builds up at the left side of the prism, as illustrated in Figure 8.10. Let us estimate the surface density of the charge: $\tilde{\sigma} = \frac{Q}{S} = \frac{\tilde{n} \Delta V e}{S} = \frac{\tilde{n} S \delta x e}{S} = e \tilde{n} \delta x$, where e is the electron charge. Electrical field produced by this excess charge is $E = \frac{\tilde{\sigma}}{\varepsilon_0} = \frac{e \tilde{n} \delta x}{\varepsilon_0}$. The modulus of the force imposed by this electrical field on the electron is $|F| = eE = \frac{e^2 \tilde{n} \delta x}{\varepsilon_0}$; the force $\vec{F}$ is directed toward the negative direction of the axis X (see Figure 8.10). The Newton Second Law for the electron now appears as follows:

$$m_e a = m_e \ddot{\delta x} = -\frac{e^2 \tilde{n} \delta x}{\varepsilon_0}. \tag{8.47}$$

This equation is easily re-written, as follows:

$$\ddot{\delta x} + \frac{e^2 \tilde{n} \delta x}{m_e \varepsilon_0} = 0. \tag{8.48}$$

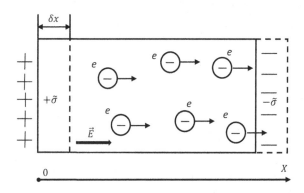

FIGURE 8.10 Origin of the plasma frequency ω_p is illustrated.

Now we compare Eq. (8.48) with Eq. (7.4), $\ddot{x} + \frac{k}{m}x = \ddot{x} + \omega^2 x = 0$; $\omega^2 = \frac{k}{m}$ (see Section 7.1). Thus, we re-write Eq. (8.48) as follows:

$$\ddot{\delta x} + \omega_p^2 \delta x = 0, \tag{8.49}$$

where ω_p, defined as "plasma frequency" is given by Eq. (8.50):

$$\omega_p = \sqrt{\frac{\tilde{n}e^2}{m_e \varepsilon_0}}. \tag{8.50}$$

Let us estimate the plasma frequency for typical metals. For example, for Li, we calculate $\tilde{n}_{Li} = 4.7 \times 10^{28}\,\mathrm{m}^{-3}$; $\omega_p^{Li} = 1.0 \times 10^{16}\,\mathrm{s}^{-1}$; for Au, we obtain $\tilde{n}_{Au} = 5.97 \times 10^{28}\,\mathrm{m}^{-3}$; $\omega_p^{Au} = 1.3 \times 10^{16}\,\mathrm{s}^{-1}$.

8.10   REFRACTION INDEX OF METALS

Now we are ready to calculate the refraction index of metals. The theory we have worked out Section 8.6 for dielectric materials can also be applied to good conductors, like metals, with very little modification. As we already mentioned in Section 8.8, when we discussed the Drude model, in metals some of the electrons have no binding force holding them to any particular atom; it is these "free" electrons which are responsible for the conductivity. There are other electrons which are bound, and the theory above is directly applicable to them. Their influence, however, is usually overlapped by the effects of the conduction electrons. We will consider now only the effects of the free electrons. If there is no restoring force on an electron but still some resistance to its motion – its equation of motion differs from Eqs. (8.20) and (8.21) only because the term in $\omega_0^2 x$ is absent. So, all we have to do is $\omega_0 = 0$ in our derivation of the refractive index of metals, except that there is one more difference. The reason that we had to distinguish between the average field and the local field in a dielectric is that in an insulator each of the dipoles is fixed in position, so that it has a definite relationship to the position of the others (see Section 8.6). But because the conduction electrons in a metal move around all over the place, the field on them on the average is just the average field E (for a detailed explanation, see Reference 6). Thus, assuming $\omega_0 = 0$ in Eq. (8.27), we calculate:

$$n^2 = 1 + \frac{\tilde{n}e^2}{m_e \varepsilon_0} \frac{1}{-\omega^2 + j\gamma\omega}. \tag{8.51}$$

Now we apply the Drude model, developed in Section 8.8. First of all, let us estimate the dissipation factor γ. The modulus of the drag force $|f_{fr}| = b\dot{x}$; consider $\beta = \frac{b}{2m_e} \rightarrow b = 2\beta m_e$. Consider $\gamma = 2\beta$ and derive $|f_{fr}| = \gamma m_e \dot{x}$. The maximal drag force is given by $|f_{fr}| = \gamma m_e v_{max}$. However, the Newton Second Law yields: $|f_{fr}| = \frac{m_e v_{max}}{\tau}$; comparing these formulas supplies (which with an accuracy of the numerical coefficient coincides with estimation made in Section 8.6):

$$\gamma = \frac{1}{\tau}. \tag{8.52}$$

Substitution of Eq. (8.2) and $\sigma = \frac{\tilde{n}e^2\tau}{2m_e}$ (see Eq. (8.44), which is one of the main results of the Drude model) into Eq. (8.51) yields the remarkable expression for the refraction index of metals:

$$n^2 = 1 + \frac{\sigma}{\varepsilon_0} \frac{1}{j\omega(1 + j\omega\tau)}. \qquad (8.53)$$

This expression demonstrates that the refraction index of metals is expressed via parameters of the Drude model, namely the conductivity of the metal and τ is the average time between two consequent collisions. We also come to the following important conclusion: (i) the refractive index of metals is given by a complex number; (ii) the refraction index of metals is frequency dependent.

8.11 REFRACTION INDEX OF METALS IN THE LOW-FREQUENCY AND HIGH-FREQUENCY APPROXIMATIONS; THE SKIN DEPTH AND THE PLASMA FREQUENCY

We already concluded that the refraction index of metals, given by Eq. (8.53) is frequency dependent. Let us see, what kind of this dependence takes place for low frequencies. Consider that in physics the notions of "low" and "high" frequencies are senseless, until the exact scale of frequencies is not exactly specified. We assume that the low frequency limit corresponds to the situation, when $\omega\tau \ll 1$ takes place, in other words $\omega \ll \frac{1}{\tau}$.

Let us estimate the numerical value of τ. From Eq. (8.44), we estimate $\tau \cong \frac{\sigma m_e}{\tilde{n}e^2}$; substituting for copper $\sigma = 5.8 \times 10^7 \, (\Omega \times m)^{-1}$; $\tilde{n} = 8.5 \times 10^{28} \, m^{-3}$; $m_e = 9.1 \times 10^{-31} \, kg$; $e = 1.6 \times 10^{-19} \, C$ we calculate $\tau \cong 2.4 \times 10^{-14} \, s$, consequently: $\omega \sim \frac{1}{\tau} \cong 10^{14} \, s^{-1}$. The time between two successive collisions of the electron may be estimated from the different reasoning, indeed, the time span between successive collisions of the electron is very roughly estimated as $\sim \frac{L}{v_{th}}$, where L is the lattice constant and v_{th} is the velocity of thermal motion of electrons (which is much larger that the velocity of the drift motion (see Exercises (8.17) and (8.20)). Assuming for copper $L \cong 3.6 \times 10^{-10} \, m$ and $v_{th} = 1.16 \times 10^5 \frac{m}{s}$ calculated for $T = 295 \, K$ (see Exercise (8.17)), yields $\tau \cong 0.3 \times 10^{-14} \, s$, which is reasonable, however, slightly lower than the value of τ emerging from actual values of thermal conductivity (for an analysis of this discrepancy, see Reference 10). Now we assume that $\omega\tau \ll 1$ takes place. When this is assumed, Eq. (8.53) yields:

$$n^2 = 1 + \frac{\sigma}{\varepsilon_0} \frac{1}{j\omega(1 + j\omega\tau)} \cong 1 + \frac{\sigma}{\varepsilon_0} \frac{1}{j\omega} \cong \frac{\sigma}{\varepsilon_0} \frac{1}{j\omega} = -j\frac{\sigma}{\varepsilon_0\omega}. \qquad (8.54)$$

Now we exploit the equation: $\sqrt{-j} = \frac{1-j}{\sqrt{2}}$ (see Exercise (8.23)). Involving this equation, we derive:

$$n = \sqrt{-j} \times \sqrt{\frac{\sigma}{\varepsilon_0\omega}} = \frac{1-j}{\sqrt{2}}\sqrt{\frac{\sigma}{\varepsilon_0\omega}} = (1-j)\sqrt{\frac{\sigma}{2\varepsilon_0\omega}} = \sqrt{\frac{\sigma}{2\varepsilon_0\omega}} - j\sqrt{\frac{\sigma}{2\varepsilon_0\omega}}, \qquad (8.55)$$

Now recall Eq. (8.31) $n = n_R - jn_j$ and compare with Eq. (8.55). We conclude:

$$n_R = n_j = \sqrt{\frac{\sigma}{2\varepsilon_0\omega}}. \tag{8.56}$$

We came to quite a surprising and very important conclusion: The real and imaginary parts of n have the same magnitude. With such a large imaginary part to n, the wave is rapidly attenuated in the metal. Recall, that the imaginary part of the refraction index $n = n_R - jn_j$ is responsible for the attenuation of the electromagnetic field in a metal (see Section 8.7, Eq. (8.33)). Let substitute $n_j = \sqrt{\frac{\sigma}{2\varepsilon_0\omega}}$ in Eq. (8.33). We recognize that the amplitude of an electromagnetic wave going in the z-direction decreases as $exp\left(-\sqrt{\frac{\sigma\omega}{2\varepsilon_0 c^2}}z\right)$ (the orientation of axes is shown in Figure 8.9, instead of the dielectric slab now we have a slab of a metallic material). Let re-write this as $exp\left(-\frac{z}{\delta}\right)$. We concluded that electromagnetic waves will penetrate into a metal only the distance δ, supplied by Eq. (8.57):

$$\delta = \sqrt{\frac{2\varepsilon_0 c^2}{\sigma\omega}}, \tag{8.57}$$

which is called in physics and materials science the *skin depth*. Let us estimate its value for a good metal, such as copper for the wavelength $\lambda = 3$ cm (which is close to those used in microwave ovens). The calculation with Eq. (8.57) yields $\delta = 6.7 \times 10^{-5} m$ (see Exercise (8.27)). We can see from this why when we exploit metallic cavities we needed to worry only about the fields inside the cavity, and not in the metal or outside the cavity. Also, we see why the losses in a cavity are reduced by a thin plating of silver or gold. The losses come from the current, which are appreciable only in a thin layer equal to the skin depth.

Now, what we have for the high frequency asymptote of refractive index in metals. Again, the notion of "high frequency" calls for explanation; the frequency is regarded as high when the condition $\omega\tau \gg 1$ is assumed. In this case, Eq. (8.53) is reduced to Eq. (8.58):

$$n^2 = 1 - \frac{\sigma/\varepsilon_0}{\omega^2\tau}. \tag{8.58}$$

For waves of high frequencies the index of a metal becomes real and it is less than one. This is a very important conclusion. Eq. (8.51) teaches us that the refraction index of metals is given by $n^2 = 1 + \frac{\tilde{n}e^2}{m_e\varepsilon_0}\frac{1}{-\omega^2 + j\gamma\omega}$. If the dissipation is neglected and $\gamma = 0$ is adopted, we obtain:

$$n^2 = 1 - \frac{\tilde{n}e^2}{\varepsilon_0 m_e\omega^2}. \tag{8.59}$$

Consider now that the plasma frequency is given by Eq. (8.50), namely: $\omega_p = \sqrt{\frac{\tilde{n}e^2}{m_e\varepsilon_0}}$. Thus, Eq. (8.58) is re-written in an extremely simple and elegant form:

$$n^2 = 1 - \frac{\omega_p^2}{\omega^2} = 1 - \left(\frac{\omega_p}{\omega}\right)^2. \tag{8.60}$$

For $\omega < \omega_P$ the index of a metal has an imaginary part, and waves are attenuated; but for $\omega \gg \omega_P$ the index is real and supplied by Eq. (8.60), and the metal becomes transparent. Metals are reasonably transparent to x-rays. But some metals are even transparent in the ultraviolet; the critical wavelength λ_p corresponding the plasma frequency ω_p is given by

$$\lambda_p = \frac{2\pi c}{\omega_p}. \tag{8.61}$$

The critical wavelength for Li is estimated as $\lambda_p = 1.55 \times 10^{-7}$ m.

8.12   REVISITING REFRACTION INDEX: REFRACTION INDEX EMERGING FROM THE MAXWELL EQUATIONS

Now we come back to the Maxwell equations summarized in Section 7.1.10. The Maxwell equations are supplied by Eqs. (8.62a,b):

$$\frac{\partial^2 \vec{E}(x,y,zt)}{\partial t^2} = \frac{c^2}{\varepsilon_r \mu_r} \nabla^2 \vec{E}(x,y,z,t) \tag{8.62a}$$

$$\frac{\partial^2 \vec{H}(x,y,zt)}{\partial t^2} = \frac{c^2}{\varepsilon_r \mu_r} \nabla^2 \vec{H}(x,y,z,t), \tag{8.62b}$$

These equations yield for the phase velocity of electromagnetic waves v_{ph}:

$$v_{ph} = \frac{c}{\sqrt{\varepsilon_r \mu_r}}, \tag{8.63}$$

where $\mu_r = \frac{\mu}{\mu_0}$ is the relative magnetic permeability of the medium, and $\varepsilon_r = \frac{\varepsilon}{\varepsilon_0}$ is the relative dielectric permittivity of substance (in vacuum $\varepsilon_r = 1$; $\mu_r = 1$). The velocity of light in vacuum is given by $c = \frac{1}{\sqrt{\varepsilon_0 \mu_0}}$ (see Section 7.1.10). Now consider Eq. (8.7) re-written as follows:

$$n = \frac{c}{v_{ph}}. \tag{8.64}$$

Substitution of Eq. (8.63) into Eq. (8.64) immediately yields Eq. (8.65):

$$n^2 = \varepsilon_r \mu_r. \tag{8.65}$$

This simple, however, very general equation serves as a basis for development of the negative refractive index (left-handed) materials to be addressed in detail in the next chapter.

Bullets
- When the incident light beam meets the slab of material a diversity of optical events take place, including reflection, refraction, absorption and scattering.
- Light reflects from a smooth surface at the same angle as it hits the surface.

- Light refraction is governed by the Snell law: $n_1 sin\theta_1 = n_2 sin\theta_2$.
- Laws of the light refraction and reflection emerge from the Fermat Principle.
- The Fermat Principle states that actual pathway of the light propagation corresponds to the local extremum of the optical path length.
- Refraction index of dielectric materials depends on the polarizability of molecules constituting the dielectric medium.
- Refraction index of dielectric material is expressed with the complex number: $n = n_R - jn_I$. The real part of the refraction index governs propagation of electromagnetic waves in a medium; whereas the imaginary part of the refractive index governs the absorption of the wave be media.
- Refraction index of metals is well explained within the Drude model, in which electrons responsible for conductivity of a metal are considered as an ideal gas.
- Plasma oscillations are oscillations of the electron density in conducting media such as metals. Consider an electrically neutral plasma in equilibrium, consisting of a gas of positively charged ions and negatively charged electrons. If one displaces a group of electrons with respect to the ions, the Coulomb force pulls the electrons back, acting as a restoring force.
- Refractive index of metals is expressed with an imaginary number.
- For low frequencies ω the real and imaginary parts of refractive index of metals have the same magnitude; $n_R = n_j = \sqrt{\frac{\sigma}{2\varepsilon_0\omega}}$, where σ is the conductivity of a metal.
- For high frequencies, i.e., for $\omega \gg \omega_p$ the refractive index is pure real and it is given by the expression: $n^2 = 1-\left(\frac{\omega_p}{\omega}\right)^2$, where $\omega_p = \sqrt{\frac{\tilde{n}e^2}{m_e\varepsilon_0}}$ is a plasma frequency, and $\tilde{n}$ is the concentration of electrons in a free electron gas.
- Generally, the refractive index of a media is given by $n^2 = \varepsilon_r\mu_r$, where μ_r is the relative magnetic permeability of the medium, and ε_r is the relative dielectric permittivity of substance (in vacuum, $\varepsilon_r = 1$; $\mu_r = 1$).

EXERCISES

1. Derive Eq. (8.1) from Eq. (7.52).
2. Derive Eq. (8.2).
   Solution: Consider plane electromagnetic wave given by $E = E_0 cos(kx - wt)$. The time-dependent energy flux across area A is given by $S(t) = \frac{energy\ passing\ area\ A\ in\ time\ \Delta t}{\Delta t} = \varepsilon_0 cE(t)^2$.
   The averaged energy of the electromagnetic wave flux is given by
   $I = c\varepsilon_0 E_0^2 \frac{1}{T} \int_0^T cos^2(\omega t)dt = c\varepsilon_0 E_0^2 \frac{1}{T} \int_0^T cos^2\left(\frac{2\pi}{T}t\right)dt = \frac{1}{2}c\varepsilon_0 E_0^2$ (compare with Eq. (8.2)).
3. Explain the meaning of the notions: reflection, retraction, absorption, scattering.
4. Derive Eq. (8.6).
   Hint: For the detailed derivation of Eq. (8.6), see Reference 4.
5. Demonstrate that for the ray entering the optical fiber from air with an incident angle θ_i (see Figure 8.5), the interrelation between the angles is given by $sin\theta_f = \frac{sin\theta_i}{n_f}$.
   Hint: Use the Snell law (Eq. (8.4)) and the fact the refraction index of air is $n = 1$.

6. Explain qualitatively propagation of light in optical fibers. Derive Eq. (8.13).
7. Formulate the Fermat Principle of least time.
8. Derive the law of reflection from the Fermat Principle of least time.
9. Derive the law of refraction (the Snell law) from the Fermat Principle of least time.
10. Define the optical path length.
11. Formulate the Fermat Principle using the notion of the optical path length.
12. Explain the notion of the stationary optical path.
13. Explain qualitatively the microscopic origin of the refraction index.
14. Derive Eq. (8.22) from Eq. (8.21).

   Hint: Substitute $x = x_0 e^{j\omega t}$ into Eq. (8.21) and consider $\gamma = 2\beta$.

15. Derive the Bouguer–Beer–Lambert law of the light absorption $\frac{I}{I_0} = e^{-\kappa z}$.

   Hint: Involve the equation of the traveling plane electromagnetic wave: $E_x = E_0 e^{j(\omega t - kz)}$ and consider: $v = \frac{\omega}{k} = \frac{c}{n}$; $n = n_R - jn_I$; $\kappa = 2\omega \frac{n_I}{c}$.

16. What is the meaning of the real and imaginary parts of the refraction index?
17. Explain the basic assumptions of the Drude Theory.
18. Estimate the thermal velocity v_{th} of electrons at a room temperature.

   Solution: According to the Drude Theory, free electrons in the metal behave as an electron gas, thus the kinetic energy of the electrons is given by $\varepsilon_k = \frac{m_e v_{th}^2}{2} = \frac{3}{2} k_B T$, where k_B is the Boltzmann constant and T is the temperature of the electron gas, hence we estimate: $v_{th} = \sqrt{\frac{3 k_B T}{m_e}}$. Assuming $m_e \cong 9.11 \times 10^{-31}\,\mathrm{kg}$; $k_B = 1.38 \times 10^{-23}\,\frac{J}{K}$; $T = 295\,\mathrm{K}$ yields $v_{th} \cong 1.16 \times 10^5\,\frac{m}{s}$.

19. Compare the thermal velocity of electrons calculated in Exercise (8.17) with the sound velocity in stainless steel.

   Solution: The sound velocity in stainless steel is $v_s \cong 5.74 \times 10^3\,\frac{m}{s}$. Thus the interrelation $v_{th} \gg v_s$; thermal velocity of electrons calculated within the assumptions of the Drude Theory is much larger than the typical velocity of sound in metals.

20. Derive Eq. (8.41).

   Solution: Consider the motion of electrons through the cylindrical conductor depicted in the sketch given in Figure 8.11. Assume that the number density of the electron gas is constant: $\tilde{n} = \frac{N}{V} = const$ and all of electrons move with the same drift velocity $\langle v \rangle$, as shown in the sketch. The cross-section of the conductor $S = const$. Electrical stream I transferred by the electrons is calculated as follows:

$$I = \frac{\Delta Q}{\Delta t} = \frac{Ne}{\Delta t} = \frac{\tilde{n} Ve}{\Delta t} = \frac{\tilde{n} \langle v \rangle \Delta t S e}{\Delta t} = \tilde{n} \langle v \rangle S e.$$

The density of the current, in turn, is given by $J = \frac{I}{S} = \tilde{n} \langle v \rangle e$. Compare with Eq. (8.40).

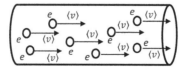

FIGURE 8.11   Motion of electrons through the cylindrical conductor is depicted.

21. Electrical current $I = 16A$ flows through ferric cylindrical conductor with a diameter $d = 0.6\,mm$.

    a. Calculate the drift velocity of electrons. Assume that the concentration of free electrons equals the concentration of atoms of Fe in the conductor (in other words, each of atoms of Fe supplies a single electron to the "ideal electron gas").

    Solution: $J = \frac{I}{S} = \tilde{n}\langle v\rangle e$, thus, $\langle v\rangle = \frac{I}{S\tilde{n}e} = \frac{4I}{\pi d^2 \tilde{n}e}$. It is necessary to calculate the concentration of free electrons $\tilde{n}$ in the ferrous conductor. $\tilde{n}$ is calculated as follows: $\tilde{n} = \frac{N_A}{\tilde{V}} = \frac{N_A}{A/\rho} = \frac{N_A\rho}{A}$, where $N_A = 6.023\times10^{23}\,mol^{-1}$, $A = 55.845\times10^{-3}\,\frac{kg}{mol}$ is the molar mass of iron and $\rho = 7.86\times10^3\,\frac{kg}{m^3}$ of iron. Substitution of the numerical values of the parameters yields the concentration of free electrons $\tilde{n} \cong 8.5\times10^{28}\,m^{-3}$. Substitution into $\langle v\rangle = \frac{4I}{\pi d^2 \tilde{n}e}$ yields $\langle v\rangle \cong 4.2\times10^{-3}\,\frac{m}{s}$. It should be emphasized that the stream density discussed in this exercise is very high (actually approaching the critical value, at which the conductor is starting melt), and correspondingly the calculated drift velocity is also to the highest possible one).

    b. compare the calculated drift velocity with the thermal velocity of electrons calculated in Exercise (8.17).

    Solution: From Exercise (8.17), we know that the thermal velocity of electrons is $v_{th} \cong 1.16\times10^5\,\frac{m}{s}$; whereas $\langle v\rangle \cong 4.2\times10^{-3}\,\frac{m}{s}$ was obtained for the drift velocity of electrons. Thus, the interrelation $v_{th} \gg v$ takes place, in other words, the velocity of thermal motion of electrons is much larger than the drift velocity of the ordered motion.

22. Demonstrate that if an electron moves with a constant acceleration a its average velocity $\langle v\rangle$ is given by $\langle v\rangle = \frac{v_{max}}{2}$, where v_{max} is the maximal velocity of electron. Assume that the initial velocity of electron is zero.

    Solution: The velocity of electron changes with time as: $v(t) = at$;, thus, for the average velocity we calculate: $\langle v\rangle = \frac{1}{\tau}\int_0^\tau v(t)\,dt = \frac{1}{\tau}\int_0^\tau at\,dt = \frac{a\tau}{2} = \frac{v_{max}}{2}$.

23. Calculate the plasma frequency for metal Na.

    Answer: $\omega_p \cong 9.0\times10^{15}\,s^{-1}$.

24. Prove: $\sqrt{-j} = \frac{1-j}{\sqrt{2}}$.

    Solution: $\left(\sqrt{-j}\right)^2 = \left(\frac{1-j}{\sqrt{2}}\right)^2$; $-j = \frac{1-2j+(j)^2}{2} = \frac{1-2j-1}{2} = -j$.

25. Demonstrate Eq. (8.55) and prove that the real and imaginary parts of n have the same magnitude: $n_R = n_j = \sqrt{\frac{\sigma}{2\varepsilon_0\omega}}$.

26. Check dimensions in equation: $n_R = n_j = \sqrt{\frac{\sigma}{2\varepsilon_0\omega}}$.

27. Consider the low-frequency limit of interaction of an electromagnetic wave with a metal (orientation of axes is shown in Figure 8.9; instead of slab of a dielectric material, we deal with a slab of metal). Demonstrate that the amplitude of an electromagnetic wave going in the z-direction decreases as $exp\left(-\sqrt{\frac{\sigma\omega}{2\varepsilon_0 c^2}}\,z\right)$.

    Hint: Substitute $n_j = \sqrt{\frac{\sigma}{2\varepsilon_0\omega}}$ into Eq. (8.33).

28. Calculate the skin depth δ for copper for the electromagnetic wave with a wavelength $\lambda = 3cm$.

    Solution: Exploit: $\delta = \sqrt{\frac{2\varepsilon_0 c^2}{\sigma\omega}}$ and $\omega = 2\pi\nu = 2\pi\frac{c}{\lambda}$. The calculation yields $\delta = 6.7\times10^{-5}m$.

29. Demonstrate that for the high frequency limit the refractive index in metals is given by $n^2 = 1 - \frac{\sigma/\varepsilon_0}{\omega^2\tau}$.

    Hint: Exploit Eq. (8.53). Assume $\omega\tau \gg 1$.

30. Derive Eq. (8.65).

REFERENCES

1. Callister W. D. Jr. *Materials Science and Engineering. An Introduction. 6th Ed.* Chapter 21, Optical Properties, John Wiley & Sons, Hoboken, NJ, USA, 2003; pp. 707–732.
2. Ashby M., Shercliff H., Cebon D. *Materials: Engineering, Science, Processing and Design, 4th Ed.*, Chapter 17, Materials for Optical Devices, Butterworth-Heinemann, Oxford, UK, 2019.
3. Fishbane P. M., Gasiorowicz S., Thornton S. T. *Physics for Scientists and Engineers*, Chapter 36, Light, Prentice Hall, Upper Saddle River, NJ, USA, 1993; Vol. 2, pp. 971–977.
4. Hecht E. *Optics. 4th Ed.*, Section 4.4 Refraction, Addison-Wesley, Reading, MA, USA, 2002; pp. 100–104.
5. Adamson P. *Philosophy in the Islamic World: A History of Philosophy without Any Gaps*, Oxford University Press, Oxford, UK, 2016; p. 77.
6. Feynman R. P., Leighton R. B., Sands M. *The Feynman Lectures on Physics*, Addison-Wesley Pub. Co, Reading, MA, 1963–1965.
7. Senior J. M., Jamro M. Y. *Optical Fiber Communications: Principles and Practice. 3rd Ed.*, Chapters 1–4, 2009, Pearson Education, Edinburgh Gate, England; pp. 12–213.
8. Cajori F. Who was the first inventor of the calculus? *Amer. Math. Monthly* 1919, **XXVI**, 15–20.
9. Veselago V. G. Formulating Fermat's principle for light traveling in negative refraction materials. *Phys.-Usp.* 2002, **45**, 1097.
10. Ashcroft N., Mermin N. D. *Solid State Physics.* Chapter 1, Holt, Rinehart and Winston, New York, 1976; pp 1–11.

9 Electromagnetic Metamaterials, Negative Refractive Index Materials

In the previous chapter, we laid the foundation for the development of optical meta-materials and introduced the notion of the refraction index of the medium. Generally speaking, the refraction index is expressed with the complex number. Media in which dielectric permittivity ε and magnetic permeability μ are negative are possible and they fulfill the demands of the Maxwell equations. In the materials, in which $\varepsilon < 0$, $\mu < 0$ take place, vectors $\vec{E}$, $\vec{H}$ and $\vec{k}$ form the left triplet; that is why these materials are called the left-handed metamaterials. In the "usual" right-handed materials, vectors of the phase velocity $\vec{v}_{ph}$ and the group velocity $\vec{v}_{gr}$ are parallel; whereas in the left- handed metamaterials, vectors $\vec{v}_{ph}$ and $\vec{v}_{gr}$ are anti-parallel. Vector of the velocity $\vec{v}_{gr}$, which indicates the direction of energy flow, is parallel to the Poynting vector $\vec{S}$; in the left-handed materials, vectors $\vec{S}$ and the wavevector $\vec{k}$ are anti-parallel. Classification of materials considering the sign of dielectric permittivity and magnetic permeability is discussed. Experimental realization of left-handed materials is discussed. Left-handed metamaterials are built from the macroscopic building blocks, contrastingly to the "usual" materials built of atoms and molecules.

9.1 NEGATIVE REFRACTIVE INDEX AND ITS ORIGIN

In the previous chapter, we discussed the origin of the phenomenon of light refraction. We established that refraction is governed by the Snell law (see Section 8.2 for details), namely $n_1 sin\theta_1 = n_2 sin\theta_2$ (see Figure 8.3), where n is the refraction index of the medium, which, being a dimensionless physical quantity, roughly speaking, gives the indication of the light-bending ability of that medium. It was Victor Georgievich Veselago in 1967 (original Russian paper was published in 1967, while the English translated paper was published in 1968) who made theoretical investigations on the solutions to Maxwell's equations in hypothetical media having simultaneously negative isotropic permittivity and permeability and observed that such material has the possibility of exhibiting the negative refractive index.[1] We follow the logic of this paper, which brought to existence the new generation of optical materials, i.e., negative refractive index materials, also known as left-handed metamaterials. In Chapter 7, we formulated the Maxwell

DOI: 10.1201/9781003178477-9

equations, the pair of which appears as follows in the situation when the current density $\vec{J} = 0$:

$$\nabla \times \vec{E} = -\frac{\partial \vec{B}}{\partial t}, \tag{9.1}$$

$$\nabla \times \vec{H} = \frac{\partial \vec{D}}{\partial t}, \tag{9.2}$$

where $\vec{E}$ is the electric field, $\vec{D}$ is the electrical displacement field (also called electrical induction, $\vec{B}$ is the magnetic field and $\vec{H}$ is the magnetic field intensity (also called magnetic field strength; see Section 7.1.10). The system of Maxwell equations is not complete until the interrelations between vectors $\vec{D}$ and $\vec{E}$ and also $\vec{B}$ and $\vec{H}$ are established. These interrelations are prescribed by Eqs. (9.3) and (9.4).

$$\vec{D} = \varepsilon_0 \varepsilon_r \vec{E} = \varepsilon \vec{E}, \tag{9.3}$$

$$\vec{B} = \mu_0 \mu_r \vec{H} = \mu \vec{H}, \tag{9.4}$$

where $\varepsilon_0 \cong 8.85 \times 10^{-12} \frac{\text{F}}{\text{m}}$ (Farads per meter) is the absolute dielectric vacuum permittivity and $\mu_0 = 4\pi \times 10^{-7} \frac{\text{H}}{\text{m}}$ (Henry per meter) is the magnetic permeability of vacuum; the constants ε_0 and μ_0 emerge in the SI system of units; ε_r and μ_r are relative electric and magnetic permittivities of the medium (ε_r is also often called the dielectric constant of the medium); $\mu = \mu_0 \mu_r$ is the magnetic permeability of the medium; $\varepsilon = \varepsilon_0 \varepsilon_r$ is the dielectric permittivity of substance (in vacuum $\varepsilon_r = 1$; $\mu_r = 1$).

Consider now the propagation of the plane electromagnetic wave, described by Eqs. (9.5) and (9.6) (see Section 7.1.10 and Figure 7.17 illustrating the plane electromagnetic wave):

$$\vec{E}(\vec{r}, t) = \vec{E}_0 e^{j(\omega t - \vec{k} \cdot \vec{r})}, \tag{9.5}$$

$$\vec{H}(\vec{r}, t) = \vec{H}_0 e^{j(\omega t - \vec{k} \cdot \vec{r})}. \tag{9.6}$$

Substitution of Eqs. (9.3) and (9.4) and Eqs. (9.5) and (9.6) into the Maxwell equations (Eqs. (9.1) and (9.2)) yields[2]:

$$\vec{k} \times \vec{E} = \omega \mu \vec{H}, \tag{9.7}$$

$$\vec{k} \times \vec{H} = -\omega \varepsilon E. \tag{9.8}$$

It is immediately recognized by Eqs. (9.7) and (9.8) that for the materials in which $\varepsilon > 0, \mu > 0$ (so called "normal" optical materials), vectors $\vec{E}$, $\vec{H}$ and wavevector $\vec{k}$ form the right triplet[2]; however, for the materials, in which $\varepsilon < 0, \mu < 0$ take place, vectors $\vec{E}$, $\vec{H}$ and $\vec{k}$ form the left triplet[2]; these materials are called the left-handed metamaterials. Conventional (right-handed) materials are often

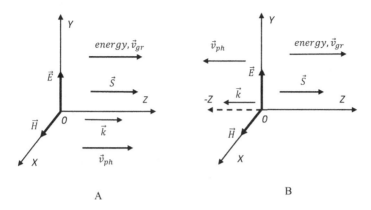

FIGURE 9.1 The vector triplet $\vec{E}$, $\vec{H}$ and $\vec{k}$ in (**A**) right-handed material (RHM) and (**B**) left-handed material (LHM). $\vec{S} = \vec{E} \times \vec{H}$ is the Poynting vector.

abbreviated RHM, whereas the left-handed materials are abbreviated LHM. The mutual orientation of the triplet of vectors $\vec{E}$, $\vec{H}$ and $\vec{k}$ in RHM and LHM is illustrated in Figure 9.1. Figure 9.1 also shows the direction of the Poynting vector $\vec{S} = \vec{E} \times \vec{H}$, introduced in Section 7.1.11, which indicates the direction of energy flow transferred by the electromagnetic wave. In RHM, vectors $\vec{k}$ and $\vec{S}$ are parallel, whereas in LHM, vectors $\vec{k}$ and $\vec{S}$ are anti-parallel (see Figure 9.1). Figure 9.1 also demonstrates the directions of phase $\vec{v}_{ph}$ and group $\vec{v}_{gr}$ velocities in RHM and LHM. The phase velocity $\vec{v}_{ph}$ is always parallel to $\vec{k}$; thus, in RHM, vector $\vec{v}_{ph}$ is parallel to vectors $\vec{k}$ and $\vec{S}$ (see Figure 9.1A), whereas, in LHM, vector $\vec{v}_{ph}$ being parallel to $\vec{k}$ is anti-parallel to vector $\vec{S}$ (see Figure 9.1B). Now consider the direction of the group velocity $\vec{v}_{gr}$, which indicates the direction of energy flow, being parallel to the Poynting vector $\vec{S}$. We conclude that in RHM, vectors $\vec{v}_{ph}$ and $\vec{v}_{gr}$ are parallel (see Figure 9.1A), whereas in LHM, vectors $\vec{v}_{ph}$ and $\vec{v}_{gr}$ are anti-parallel (see Figure 9.1B).

Now we address the value of the refractive index. Recall Eq. (7.86) from Section 7.1.10:

$$v_{ph} = \frac{c}{\sqrt{\varepsilon_r \mu_r}}, \tag{9.9}$$

and compare Eq. (9.9) to Eq. (8.7) from Section 8.2:

$$n_r = \frac{c}{v_{ph}}; \ v_{ph} = \frac{c}{n_r}. \tag{9.10}$$

Comparing Eq. (9.9) to Eq. (9.10) immediately yields:

$$n_r^2 = \varepsilon_r \mu_r. \tag{9.11}$$

It is immediately seen from Eq. (9.11) that simultaneous change in the sign of ε_r and μ_r, i.e., $\varepsilon_r < 0$, $\mu_r < 0$ does not change the value of the relative refractive index

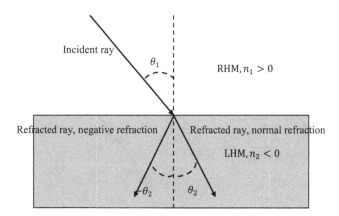

FIGURE 9.2 The light comes from the RHM to the LHM medium. The light beam is bent in the direction $-\theta$. $|n_2| > |n_1|$.

n_r as it was pointed out by Veselago in Reference 1; moreover, n_r may be negative (we are already well accustomed with the complex refractive index from the previous chapter). We already mentioned in Section 8.2 that the refractive index (or refraction index) gives the indication of the light-bending ability of that medium. Consider the situation when the light comes from the RHM medium (with the positive refractive index) to the LHM medium (with the negative refractive index) as shown in Figure 9.2; the refracted ray is bent in the $-\theta$ direction and we say that the LHM medium exhibits a negative refractive index. The situation depicted in Figure 9.2 implies $|n_2| > |n_1|$.

Now we address one of the most amazing effects, which became possible due to metamaterials. We mean development of super-lenses.[4] Consider first the ordinary conventional lens depicted in Figure 9.3A. From geometric optics, we expect the rays from the source to be focused at a point.[5] However, if we consider the wave nature of electromagnetic radiation, several Fourier components of the wave are superimposed at the focal point and the maximum resolution of the image, denoted Δ, can never be greater than approximately $\frac{\lambda}{2}$, where λ is the wavelength of the light (according to Rayleigh criterion, two images are considered as resolvable when the center of the diffraction pattern of one is directly over the first minimum of the diffraction pattern of the other).[6] In a very paradoxical way, the rectangular prism/slab with a thickness of d made from the LHM will work as a lens as demonstrated in Reference 4 and depicted in Figure 9.3B.

Indeed, the light produced by the point source is focused by slab in a point (see Figure 9.3B). Moreover, this lens works as a "perfect lens"; in other words, its resolution limit Δ is different from that of the "usual lens". Consider the LHM slab with a thickness of d (see Figure 9.3B). For the lenses made from the metamaterial, the resolution limit as a rule of thumb is approximately $2d$.[7] More accurate analysis is supplied in Reference 8. Consider the LHM, in which $\varepsilon_r = -1 + j\varepsilon_I$; $\mu_r = -1 + j\mu_I$ takes place. The maximal resolution of the image according to Reference 8 is given by Eq. (9.12):

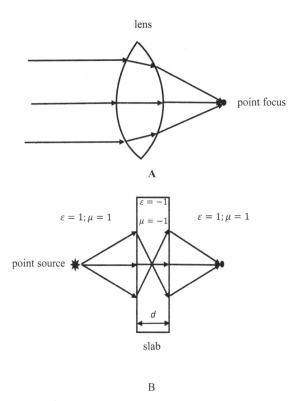

FIGURE 9.3 Comparison of the ordinary lens (**A**) vs. superlens, shown in inset (**B**). Ordinary lens focuses the light in a point/focus. A rectangular slab with the thickness of d made of LHM, shown in inset **B,** focuses light produced by the point source.

$$\Delta \cong \frac{2\pi d}{\ln \dfrac{\varepsilon_I}{2}}. \tag{9.12}$$

When the thickness of the slab made from the LHM metamaterial is smaller than the quarter wavelength $\Delta < \frac{\lambda}{4}$, the resolution of the slab lens made of metamaterial is better than the resolution of the lens made from the usual optical material.

9.2 OPTICAL CLOAKING WITH METAMATERIALS

The idea of the cloaking of a proscribed volume of space to exclude completely all electromagnetic fields is presented in Reference 9. The idea suggested in Reference 9 is to achieve concealment by cloaking the object with a metamaterial whose function is to deflect the rays that would have struck the object, guide them around the object and return them to their original trajectory.[9] For simplicity, the authors choose the hidden object to be a sphere of radius R_1 and the cloaking region to be contained within the annulus $R_1 < r < R_2$, as it is depicted in

FIGURE 9.4   Electromagnetic cloaking with metamaterials is presented.

Figure 9.4. It is shown in Reference 9 that matching of the electromagnetic characteristics of the cloaked object and surrounding material enables to exclude all fields from the central region.

9.3   DOUBLE-NEGATIVE METAMATERIALS, HOW THEY ARE BUILT?

The very question is: how to develop and manufacture the LHM material? One of the simplest ways to do it was suggested by Sir John Brian Pendry in the series of papers (see References 10 and 11; for the first reading, Reference 11 is strongly recommended). A metamaterial with the negative dielectric permittivity $\varepsilon_r < 0$ is a periodic structure, built of thin infinite metallic wires, arranged in a simple cubic lattice with the constant a depicted in Figure 9.5.

This structure mimics the electromagnetic response of plasma (see Sections 7.1.9 and 8.9). When plasma is exposed to harmonic electromagnetic waves, oscillations of the charge density occur, which take place with the plasma frequency, $\omega_p = \sqrt{\frac{\tilde{n}e^2}{m_e \varepsilon_0}}$. We calculated the plasma frequency for typical metals in Section 8.9 and demonstrated that it is located for these metals in the UV-band of the spectrum. We already calculated the refraction index for metals in the previous chapter (recall Eq. (8.60): $n^2 = 1 - \frac{\omega_p^2}{\omega^2} = 1 - \left(\frac{\omega_p}{\omega}\right)^2$ and $\varepsilon_r = n^2$); thus, dielectric permittivity ε_r, in turn, is given by Eq. (9.13):

$$\varepsilon_r = 1 - \frac{\omega_p^2}{\omega^2}. \tag{9.13}$$

The question is how to calculate the plasma frequency for the system shown in Figure 9.5? In the metallic wire system depicted in Figure 9.5, the electron density is much reduced relative to solid metal. Thus, we necessarily introduce the "effective"

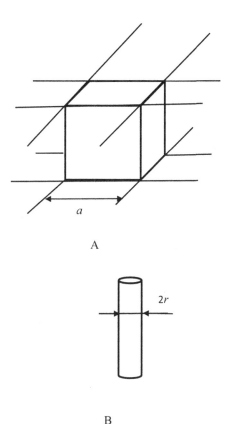

A

B

FIGURE 9.5   Metamaterial with $\varepsilon < 0$ is depicted. (**A**) Metamaterial is a periodic structure composed of thin infinite metallic wires, arranged in a simple cubic lattice with the constant a. The lattice mimics the response of plasma. In a typical example, the wires might be a few tens of microns in diameter and spaced by a few millimeters, giving a plasma frequency in the GHz range. (**B**) Metallic wire is shown.

density of electrons, denoted n_{eff}. Furthermore, there is an additional contribution to the effective mass of the electrons m_{eff} emerging from the inductance of the wires. The plasma frequency in the addressed system is calculated with Formulas (9.14)–(9.16):

$$\omega_p^2 = \frac{n_{eff}e^2}{\varepsilon_0 m_{eff}}, \tag{9.14}$$

where

$$n_{eff} = \tilde{n}\frac{\pi r^2}{a^2}, \tag{9.15}$$

and

$$m_{eff} = \frac{\mu_0 e^2 r^2 \tilde{n}}{2\pi} \ln\frac{a}{r}, \tag{9.16}$$

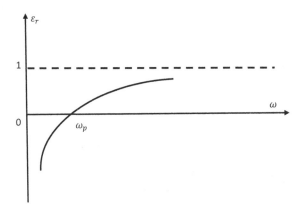

FIGURE 9.6 The dependence $\varepsilon_r(\omega)$ for plasmas. Below the plasma frequency ω_p, the dielectric permittivity $\varepsilon_r(\omega)$ is negative.

where $\tilde{n}$ is the charge-carrier number density in the metal wire, a is the lattice constant and r is the radius of the wire (see Figure 9.5B). Typical dimensions of a few microns radius for the wires and a spacing of a few millimeters give effective masses to the electrons of the same order as a nitrogen atom and as a result the plasma frequency is depressed into the GHz region (see Exercises 9.1–9.3). Note how this very radical change in dielectric properties is brought about by an extremely small amount of metal. In our example, the concentration of metal is only a few parts per million, comparable with the levels of dopants in a semiconductor.

It is easily seen from Eq. (9.13) that below the plasma frequency ω_p the dielectric permittivity $\varepsilon_r(\omega)$ is negative, as illustrated with Figure 9.6.

Now we propose the following classification of materials,[12] as they are seen from the point of view of their electromagnetic properties and as it is illustrated in Figure 9.7:

  i. Regular or common materials, for which $\varepsilon_r > 0; \mu_r > 0$ is true. These are common dielectric materials, such as polymers and ceramics.
 ii. Mu-negative materials for which $\varepsilon_r > 0; \mu_r < 0$ takes place. This kind of materials usually exploits ferrites.[13,14]
iii. Double-negative, LHM materials (see References 7–11, and Figure 9.1), in this case: $\varepsilon_r < 0; \mu_r < 0$ is true.
 iv. Epsilon-negative materials, which include metals, plasmas; $\varepsilon_r < 0; \mu_r > 0$ is true for these materials.

9.4 MAGNETIC METAMATERIALS

For materials at optical frequencies, the dielectric permittivity ε_r is generally different from that in vacuum. For example, negative permittivity values, as seen from Eq. (9.13), are routinely observed in noble metals at frequencies less than their plasma frequency. In contrast, the magnetic permeability for naturally occurring materials is

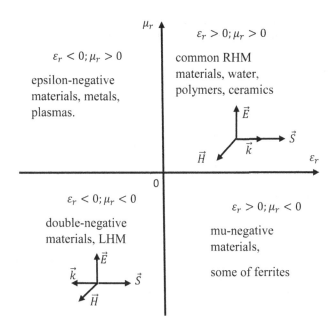

FIGURE 9.7 Classification of materials in a view of their electromagnetic properties is presented.

always close to its free space value in the optical range. The nearly absence of a magnetic response appears as a general rule in the field of optics. In the classical textbook by Landau and Lifshitz,[15] there is a general statement, explaining the reason why a magnetic response resulting from orbital currents in atoms should be negligible at optical frequencies, and thus, "the magnetic permeability ceases to have any physical meaning at relatively low frequencies... There is certainly no meaning in using the magnetic susceptibility from optical frequencies onwards, and in discussion of such phenomena we must put $\mu_r = 1$".[15] In other words, the diamagnetism is a weak phenomenon; indeed, the most strongly diamagnetic material is bismuth, with the magnetic susceptibility $\chi_V = -1.66 \times 10^{-4}$ (as expressed with SI Units).

The major reason behind the absence of optical magnetism in nature is that the magnetic field component of light couples to atoms much more weakly than the electric component, making light interacting with matter a "one-handed" situation.[16] The magnetic coupling to an atom is proportional to the Bohr magneton Bohr μ_B, defined according to $\mu_B = \frac{e\hbar}{2m_e c} = \frac{\alpha e a_0}{2}$, while the electric coupling is ea_0, where $a_0 = \frac{\hbar}{m_e c \alpha}$, and $\alpha = \frac{e^2}{4\pi\varepsilon_0\hbar c} \cong \frac{1}{137}$ is the fine structure constant. The induced magnetic dipole also contains the fine structure constant $\alpha \cong \frac{1}{137}$, so the effect of light on the magnetic permeability is two times weaker than light's effect on the electric permittivity. This means that of the two field components of light – electric and magnetic – only the electric "hand" efficiently probes the atoms of a material, while the magnetic component remains relatively unused.[16] Consequently, in all conventional optical materials and devices, only the electric component of light is directly controlled. The magnetic field component of light plays merely an auxiliary role through its relation with the electric field governed by Maxwell's equations.[15,16] It should also be emphasized that there are no free magnetic monopoles, and thus it is impossible to prepare a magnetic plasma as we can do with electrons. Therefore, it is indeed a challenging issue to achieve any magnetic response in the microwave frequencies and higher, enabling negative magnetic permeability.[16]

Recently, however, the emergence of magnetic metamaterials has fundamentally altered the situation. In metamaterials that consist of artificial subwavelength structures with tailored properties, the magnetic response is not limited anymore to the electronic spin states of individual atoms. Instead, magnetism can be achieved even in optical frequencies by specially designed "meta-atoms" – functional, macroscopic units of the metamaterial that are smaller than the wavelength.

One of these functional units is split-ring resonator, abbreviated SRR and depicted in Figure 9.8.

Let us explain how SRR works.[16–18] First, let us consider a circular metal plate placed in an oscillating electromagnetic wave with the magnetic field oriented normally to the flat surface (we follow Reference 16 in our analysis). Is the metal plate magnetically active in this case? The answer is yes, but the magnetic response is weak. The oscillating magnetic field induces a circular current in the round plate, which produces a magnetic flux opposing the external magnetic field. This can be

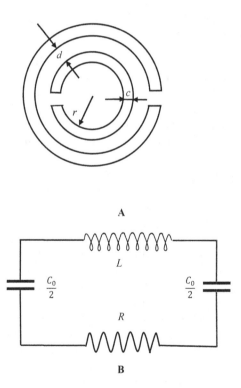

FIGURE 9.8 (**A**) Split-ring resonator (SRR) is depicted. (**B**) Equivalent electrical scheme of SRR is shown.

viewed as a simple result of Lenz's law: any induced electromotive force will be in the direction such that the flux it creates will oppose the change in the flux that produces it. Consequently, a metal plate is weakly diamagnetic and an array of such plates would exhibit an effective permeability slightly less than unity.[16] Because the circular current is mostly confined to the outer perimeter, we can remove the inner part of the plate and the plate evolves into a ring.[16-18] The response of a metallic ring to the external magnetic field is purely inductive and non-resonant. To introduce a resonance behavior and enhance the magnetic response, capacitance can be purposely introduced. As seen in Figure 9.8, a gap in each metallic ring prevents the formation of a complete circular current and charges accumulate across the gaps. With both capacitance C and inductance L, the SRR is a resonant element.[16-18] Capacitance is more effectively introduced when two rings are placed concentrically with their gaps opposite of each other (see Figure 9.8A), which is one reason why double SRRs are preferable to single SRRs in metamaterial designs. The rigorous calculation of the effective magnetic permeability of the SRR element yields (see Reference 16):

$$\mu_{effr} = 1 + \frac{m_H}{VH},\tag{9.17}$$

where H is the external magnetic field, V is the volume of each unit SRR cell and m_H is the magnetic moment of the unit SRR cell, which is calculated as follows:

$$m_H = \frac{\mu_0 \pi^2 r^4}{L\left(\dfrac{\omega_0^2}{\omega^2} - 1\right)},\tag{9.18}$$

where frequency ω_0 is supplied by Eq. (9.19):

$$\omega_0 = \sqrt{\frac{2}{\left(L + \dfrac{R}{j\omega_0}\right)C}},\tag{9.19}$$

where $C = \frac{C_0}{4}$; $C_0 = \frac{2\pi r \varepsilon_0 (c+t)}{d}$; $R = \frac{\pi r}{c\delta\sigma}$; $L \cong 2\mu_0 r$, where t is the thickness of the metal film, r, c, d are geometrical parameters of SRR, shown in Figure 9.8A, with δ and σ being the skin depth and the conductivity of the metal, respectively.[16] The equivalent electrical scheme of SRR is shown in Figure 9.8B.

From these equations, we learn that both dia- ($\mu_{effr} < 1$) or paramagnetic ($\mu_{effr} > 1$) responses to the external magnetic field of SRR are possible, depending on whether the wavelength of the incoming magnetic field is shorter or longer than the resonance wavelength. The effective permeability is therefore different from unity, even though the natural materials comprising the SRR array have unity permeability at frequencies at about a gigahertz or higher. Again, we note that this is a very rough analysis; a more accurate analysis is supplied in Reference 16. SRR opens the way to the development of metamaterials with the effective magnetic permeability. In these materials, stacks of SRR are used for controlling the effective magnetic response of the metamaterial.[6]

9.5 DEVELOPMENT AND MANUFACTURING OF THE DOUBLE-NEGATIVE METAMATERIALS

It seems that the first practical realization of LHM was reported in Reference 19. The LHM reported in Reference 19 consists of a two-dimensional periodic array of copper split-ring resonators and wires, fabricated by a shadow mask/etching technique on 0.25-mm thick G10 fiber glass circuit board material. Thus, the reported metamaterial combines the ideas illustrated with Figures 9.5 and 9.8 giving rise to the negative values of ε_r and μ_r. After processing, the boards were cut and assembled into an interlocking unit, from which a prism-shaped section was cut for the beam-deflection experiments. In the reported metamaterial, ε_r and μ_r are both negative about 10.2–10.8 GHz.[19] The refraction index is also expected to take very negative values on the low-frequency side of the left-handed band, passing through a value of zero on the high-frequency side.[19]

An alternative approach to practical realization of LHM was suggested in Reference 20. In Reference 20, metamaterial that exploits the well-known L-C distributed network representation of homogeneous dielectrics was reported. In the conventional low-pass topology, the quantities L and C represent a positive equivalent permeability and permittivity, respectively.[20] However, in the dual configuration, in which the positions of L and C are simply interchanged, these equivalent material parameters assume simultaneously negative values.[20] Two-dimensional periodic versions of these dual networks were used to demonstrate negative refraction and focusing, discussed in Section 9.1. The review of modern methods, suggested for manufacturing LHM, is supplied in References 3 and 16.

9.6 NEGATIVE REFRACTION INDEX AND THE FERMAT PRINCIPLE

In Section 8.5, we formulated the least time Fermat principle. The Fermat principle states that the actual path between two points taken by a beam of light is the one as traversed in the least time. The aforementioned interpretation of the Fermat principle does not work for LHM.[21] The correct, accurate wording of the Fermat principle which is true for both "usual/regular" and metamaterials addresses the optical path length L defined with Eq. (9.19)

$$L = \int_C n\, ds, \qquad (9.20)$$

where n is the local refractive index as a function of distance along the path C.[22] The correct formulation of the Fermat principle which works for both LHM and RHN sounds as follows: the optical path length of the actual ray, given by Eq. (9.19), is either an extremum (a minimum or a maximum) with respect to the optical path length of adjacent paths or equal to the optical path length of adjacent paths.[21]

Bullets

- Media with negative values dielectric permittivity ε and magnetic permeability μ are possible and they fulfill the demands of the Maxwell equations.

- In the materials, in which $\varepsilon < 0, \mu < 0$ takes place, vectors $\vec{E}, \vec{H}$ and $\vec{k}$ form the left triplet.
- These materials are called the left-handed metamaterials.
- In the usual/regular right-handed materials, vectors of the phase velocity $\vec{v}_{ph}$ and the group velocity $\vec{v}_{gr}$ are parallel.
- In the left-handed metamaterials, vectors $\vec{v}_{ph}$ and $\vec{v}_{gr}$ are anti-parallel; vectors $\vec{S}$ and $\vec{k}$ are also anti-parallel. Experimental realization of left-handed materials is discussed.
- Left-handed metamaterials are built from the macroscopic building blocks, contrastingly to the "usual" materials built of atoms and molecules.
- Periodic structure, built of thin infinite metallic wires, arranged in a simple cubic lattice provides in a certain range of frequencies the negative value of the dielectric permittivity.
- Split-ring resonators in a certain range of frequencies provide negative values of magnetic permeability.
- Left-handed metamaterials make possible "perfect lenses" and electromagnetic cloaking of physical bodies.

EXERCISES

1. Consider the meta-structure, depicted in Figure 9.5A, built of the gold wires with a radius of $r = 1.0$ μm; the lattice constant of the structure is $a = 1$ mm. The concentration of free electrons in gold is $\tilde{n}_{Au} = 5.97 \times 10^{28} \, \mathrm{m}^{-3}$. Calculate the effective mass of electron in the lattice. Compare it to the mass of the atom of Hydrogen.
    Solution: The effective mass of electron is given by $m_{eff} = \frac{\mu_0 e^2 r^2 \tilde{n}}{2\pi} \ln \frac{a}{r}$, considering $\mu_0 = 4\pi \times 10^{-19} \frac{H}{m}; e = 1.6 \times 10^{-19} C$ and substitution of the numerical data yields: $m_{eff} \cong 2.11 \times 10^{-27}$ kg which is comparable to the mass of the Hydrogen atom which is $m_H = 1.6 \times 10^{-27}$ kg.

2. Calculate the effective density of electrons n_{eff} in the lattice depicted in Figure 9.5A, built of the gold wires with a radius of $r = 1.0$ μm; the lattice constant of the structure is $a = 1$ mm.
    Solution: The effective density of electrons is calculated with Eq. (9.15):
    $n_{eff} = \tilde{n} \frac{\pi r^2}{a^2}$.
    Substitution $\tilde{n}_{Au} = 5.97 \times 10^{28} \, \mathrm{m}^{-3}$ yields: $n_{eff} \cong 18.7 \times 10^{22} \, \mathrm{m}^3$.

3. Calculate the plasma frequency for the aforementioned lattice depicted in Figure 9.5A. Compare the calculated value with the plasma frequency established for the free electrons in gold.
    Solution: The plasma frequency is calculated with Eq. (9.15): $\omega_p^2 = \frac{n_{eff} e^2}{\varepsilon_0 m_{eff}}$. The effective mass of the electron and the effective density were established in the previous exercises as $m_{eff} \cong 2.11 \times 10^{-27}$ kg, and $n_{eff} \cong 18.7 \times 10^{22} \, \mathrm{m}^3$. Substitution of the numerical data yields: $\omega_p \cong 505 \; \mathrm{GHz} = 5.05 \times 10^{11} \mathrm{Hz}$, which is much lower than the plasma frequency established for free electrons in gold, namely: $\omega_p^{Au} = 1.3 \times 10^{16} \mathrm{s}^{-1}$ (see Section 8.9).

REFERENCES

1. Veselago V. G. The electrodynamics of substances with simultaneously negative values of ε and μ. *Sov. Phys. Usp.* 1968, **10**, 509–514.
2. Arfken G. B., Weber H. J. *Mathematical Methods for Physicists, 5th Ed.*, A Harcourt Science and Technology Company, San Diego, USA, 2001.
3. Kar S. Metamaterials or Left-Handed Materials - A Counter-Intuitive Artificial Material, Chapter 1 in *Metamaterials and Metasurfaces, Basics and Trends*; IOP Publishing Ltd, Bristol, UK, 2023; pp. 1–46.
4. Pendry J. B. Negative refraction makes a perfect lens. *Phys. Rev. Lett.* 2000, **85** (18), 3966–3969.
5. Hecht E. Optics, Chapter 5 in *Geometrical Optics, 4th Ed.*, Addison-Wesley, Reading, MA, USA, 2002; pp. 149–231.
6. Hecht E. Optics, Chapter 9 in *Geometrical Optics, 4th Ed.*, Addison-Wesley, Reading, MA, USA, 2002; p. 424.
7. Fang N., Zhang X. Imaging properties of a metamaterial superlens. *Appl. Phys. Lett.* 2003, **82**, 161.
8. Podolskiy V. A., Kuhta N. A., Milton G. W. Optimizing the superlens: Manipulating geometry to enhance the resolution. *Appl. Phys. Lett.* 2005, **87**, 231113.
9. Pendry J. B., Schurig D., Smith D. R. Controlling electromagnetic fields. *Science* 2006, **312**, 1780–1782.
10. Pendry J. B., Holden A. J., Stewart W. J., Youngs I. Extremely low frequency plasmons in metallic mesostructures. *Phys. Rev. Lett.* 1996, **76**, 4773.
11. Pendry J. B. Negative refraction. *Contemp. Phys.* 2004, **45** (3), 191–202.
12. Fan G., Sun K., Hou Q., Wang Z., Liu Y., Fan R. Epsilon-negative media from the viewpoint of materials science. *EPJ Appl. Metamat.* 2021, **8**, 11.
13. Thummaluru S. R., Chaudhary R. K. Mu-negative metamaterial filter-based isolation technique for MIMO antennas. *Electron. Lett.* 2017, **53** (10), 644–646.
14. Bi K., Zhou J., Liu X., Lan C., Zhao H. Multi-band negative refractive index in ferrite-based metamaterials. *Progr. Electromag. Res.* 2013, **140**, 457–469.
15. Landau L. D., Liftshitz E. M., Pitaevskii L. P. *Electrodynamics of Continuous Media*, Chapter 79, Pergamon, New York, USA, 1984.
16. Cai W., Shalaev V. *Optical Metamaterials, Fundamentals and Applications*, Chapter 5, Springer, New York, 2010.
17. Pendry J. B., Holden A. J., Robbins D. J., Stewart W. J. Magnetism from conductors and enhanced nonlinear phenomena. *IEEE Trans. Microw. Theory Tech.* 1999, **47**, 2075–2084.
18. Hardy W. N., Whitehead L. A. Split-ring resonator for use in magnetic-resonance from 200–2000 MHz. *Rev. Sci. Instrum.* 1981, **52**, 213–216.
19. Shelby R. A., Smith D. R., Schultz S. Experimental verification of a negative index of refraction. *Science* 2001, **292**, 77–79.
20. Eleftheriades G. V., Iyer A. K., Kremer P. C. Planar negative refractive index media using periodically L-C loaded transmission lines. *IEEE Trans. Microw. Theory Tech.* 2002, **50**, 2702–2712.
21. Veselago V. G. Formulating Fermat's principle for light traveling in negative refraction materials. *Phys.-Usp.* 2002, **45**, 1097.
22. Hecht E. Optics, Section 4.4, Refraction, 4th Ed., Addison-Wesley, Reading, MA, USA, 2002; pp. 100–104.

10 The Effect of Negative Mass and Negative Density
Acoustic Metamaterials

In the previous chapter, we discussed electromagnetic metamaterials, exploiting the effect of the negative refractive index. Now, we address acoustic metamaterials enabling effective manipulation of sound waves. The notions of negative mass and negative density are introduced and discussed. The effects of the negative mass and negative density emerge from the oscillations in the vicinity of the resonance frequency. Mechanical models giving rise to the negative effective mass and negative effective density are introduced. Metamaterials built of the units featured by the negative mass are introduced. Auxetic materials characterized by the negative value of the bulk modulus are addressed. Double-negative acoustic materials demonstrating simultaneous negativity in density and bulk modulus are discussed. Applications of the acoustic metamaterials are addressed.

10.1 THE EFFECT OF NEGATIVE MASS

In Chapter 7, we demonstrated that the propagation of both electromagnetic and acoustic waves is described by the same wave differential equation (see Sections 7.1.8 and 7.1.10). Processes governed by the same equations are well expected to demonstrate similar physical features. Thus, we expect that the negative refraction of acoustic waves will be possible in the meta-media. In order to understand how negative refraction of acoustic waves is realized with metamaterials, we have to acquaint ourselves with the notions of the negative mass and negative density. The concept of the negative mass was suggested in References 1–3 and mathematically carefully developed in Reference 4. Consider first the one-dimensional model depicted in Figure 10.1, giving rise to the phenomenon of the negative mass.[4] In this 1D model, n cylindrical cavities of length d have are imbedded into a bar of rigid material.[4] In the center of this hollow chamber cavity, a lead sphere of mass m and radius r is located. The lead sphere is attached to the ends of the cavity with two (generally speaking, viscoelastic) springs, each having the same complex spring constant k (now the concept of the complex numbers, introduced into Section 7.1.2, will be of a primary importance). For the sake of simplicity, we neglect gravity and assume the springs are un-stretched in their equilibrium configuration, i.e., when they have length $\frac{d}{2} - r$. We assume that the entire system moves harmonically with time with angular frequency ω. The spring constant k may depend on the frequency. Assume that the harmonic

DOI: 10.1201/9781003178477-10

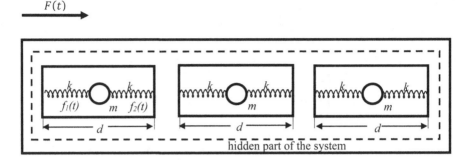

FIGURE 10.1 A one-dimensional metamaterial where the effective mass depends on the frequency ω and can be negative.

external force acting on the entire system is $F(t)$; the force $f_1(t)$, the spring on the left of each mass exerts on the bar, and the force $f_2(t)$, the spring on the right of each mass exerts on the bar, are given by (see Reference 4 and Section 7.1.2):

$$F(t) = Re\left(\hat{F}e^{-j\omega t}\right); \; f_1(t) = Re\left(\hat{f}_1 e^{-j\omega t}\right); f_2(t) = Re\left(\hat{f}_2 e^{-j\omega t}\right), \qquad (10.1)$$

where $\hat{F}$, $\hat{f}_1$ and $\hat{f}_2$ are supplied by the complex numbers. According to Newton's Second Law, we have:

$$F(t) = \frac{dp(t)}{dt}, \text{ where } p(t) = Re\left(\hat{p}e^{-j\omega t}\right), \qquad (10.2)$$

in which $p(t)$ is the momentum of the entire system and $\hat{p}$ is given by the complex number. In the first cavity/chamber, the left wall is at $X(t)$ and the center of the lead sphere is at $x(t)$, given by Eq. (10.3):

$$X(t) = Re\left(\hat{U}e^{-j\omega t}\right), \; x(t) = \frac{d}{2} + Re\left(\hat{u}e^{-j\omega t}\right), \qquad (10.3)$$

where $\hat{U}$ and $\hat{u}$, appearing in Eq. (10.3), are the complex displacements of the rigid bar and each lead ball, respectively. The velocity of the rigid bar $\frac{dX(t)}{dt}$ is given by Eq. (10.4):

$$\frac{dX(t)}{dt} = Re\left(\hat{V}e^{-j\omega t}\right), \text{ where } \hat{V} = -j\omega\hat{U}, \qquad (10.4)$$

while the velocity of each lead sphere m is given by Eq. (10.5):

$$\frac{dx(t)}{dt} = Re\left(\hat{v}e^{-j\omega t}\right), \text{ where } \hat{v} = -j\omega\hat{u}, \qquad (10.5)$$

Assuming the rigid bar has a mass M_0, and applying Newton's Second Law of motion to each lead sphere yields Eq. (10.6):

$$\hat{F} = -j\omega M_0 \hat{V} + n\left(\hat{f}_1 - \hat{f}_2\right) = -j\omega\left(M_0\hat{V} + nm\hat{v}\right) \qquad (10.6)$$

Recall that n is the number of cavities. Eq. (10.6) may be re-written as follows:

$$\hat{F} = -j\omega\hat{p}, \text{ where } \hat{p} = M_0\hat{V} + nm\hat{v}. \tag{10.7}$$

It should be emphasized that velocity $\hat{v}$ is unobservable since it is in the hidden part of the bar, as depicted in Figure 10.1 (the hidden part of the system is encircled within the dashed line). Thus, we need to relate the complex observable momentum $\hat{p}$ to the complex observable velocity of the entire bar $\hat{V}$. Hooke's law for each spring implies for the complex elastic forces the following:

$$\hat{f}_1 = k\left(\hat{U} - \hat{u}\right) = -\hat{f}_2. \tag{10.8}$$

Substituting Eq. (10.8) into Eq. (10.6) yields:

$$2k\left(\hat{U} - \hat{u}\right) = -m\omega^2\hat{u}. \tag{10.9}$$

Solution of Eq. (10.9) is supplied by Eq. (10.10):

$$\hat{u} = \frac{2k}{2k - m\omega^2}\hat{U}. \tag{10.10}$$

Consider that $\lim_{k\to\infty}\hat{u} = \hat{U}\lim_{k\to\infty}\frac{2k}{2k-m\omega^2} = \hat{U}$. Thus, in the case of infinitely stiff springs $(k \to \infty)$, $\hat{u} \cong \hat{U}$. Finally, we obtain Eq. (10.11):

$$\hat{p} = m_{eff}\hat{V}; \; m_{eff} = M_0 + \frac{2kn}{2k - m\omega^2}, \tag{10.11}$$

where $m_{eff} = M_0 + \frac{2kn}{2k-m\omega^2}$ is called the "effective mass" of the system, shown in Figure 10.1. The effective mass m_{eff} may be complex and huge near the resonance (see Section 7.1.6), namely when $\omega^2 \cong \frac{2k}{m}$ (consider that friction is absent within the addressed system). Moreover, the real part of the effective mass may be negative, when the frequency ω approaches the resonance value $\omega = \sqrt{\frac{2k}{m}}$ from above. The negative effective mass means that the force acting on the bar depicted in Figure 10.1 and the acceleration of the bar are oriented in the opposite directions. The effect becomes possible due to the oscillations of the hidden masses m (see Figure 10.2) and it is the pronounced nature resonance effect, observed only close to the fixed resonance frequencies.

The alternative mechanical model giving rise to the negative effective mass effect is depicted in Figure 10.2. Now, we consider a single unit built of the core, shell and elastic massless spring, as depicted in Figure 10.2. A core with mass m_2 is connected internally through the spring with k_2 to a shell with mass m_1. The system is subjected to the external sinusoidal force $F(t) = F_0\sin\omega t$. If we solve the equations of motion for the masses m_1 and m_2 and replace the entire system with a single effective mass m_{eff} in the aforementioned way, we eventually obtain:

$$m_{eff} = m_1 + \frac{m_2\omega_0^2}{\omega_0^2 - \omega^2}, \tag{10.12}$$

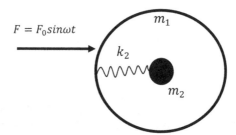

FIGURE 10.2 Core-shell system giving rise to the effect of the negative mass. A core body with mass m_2 is connected internally through the spring with k_2 to a shell with mass m_1.

where $\omega_0 = \sqrt{\frac{k_2}{m_2}}$. Obviously, when the frequency ω approaches ω_0 from above, the effective mass m_{eff} supplied by Eq. (10.12) will be negative.[1-3] Again negative effective mass emerging from Eq. (10.12) reflects the fact that the external sinusoidal force $F(t) = F_0 sin\omega t$ and the acceleration of the shell are oriented in the opposite directions.

10.2 THE EFFECT OF NEGATIVE MASS AND PLASMA OSCILLATIONS

The elastic spring appearing in Figure 10.2 should not necessarily be a mechanic spring. The spring k_2 may represent the physical process, giving rise to harmonic oscillations. Such a process may be the plasma oscillations of a free electron gas, addressed in detail in Section 8.9. We consider the electro-mechanical analogy of the aforementioned mechanical model, shown in Figure 10.2, giving rise to the negative effective mass.[5] Consider cubic metal particle, seen as ionic lattice m_1 containing the Drude-Lorenz free electrons gas (see Section 8.8). The electron gas possesses the total mass of $m_2 = m_e \tilde{n} V$, where $m_e = 9.1 \times 10^{-31}$ kg is the mass of electron, $\tilde{n}$ is the charge-carrier number density $\tilde{n} = \frac{N}{V}$, where N is the total number of charge carriers/electrons in a volume V, $[\tilde{n}] = m^{-3}$. Electron gas is free to oscillate with the plasma frequency $\omega_p = \sqrt{\frac{\tilde{n} e^2}{m_e \varepsilon_0}}$ (see Section 8.9).

Consider the metallic particle exerted to the external sinusoidal force $F = F_0 sin\omega t$. The effective mechanical scheme of the metallic particle is shown in Figure 10.3B (the right sketch) and it exactly coincides with that giving rise to the negative effective mass, supplied in this case by:

$$m_{eff} = m_1 + \frac{m_2 \omega_p^2}{\omega_p^2 - \omega^2}, \tag{10.13}$$

where m_1 is the mass of the ionic lattice, m_2 is the total mass of the electronic gas and $k_2 = \omega_p^2 m_2$; it is seen that it may be negative when the frequency ω approaches ω_p from above. Considering $\frac{m_2}{m_1} \ll 1$ yields:

$$\frac{m_{eff}}{m_1 + m_2} \cong \frac{m_{eff}}{m_1} \cong 1 + \frac{m_2}{m_1} \frac{\omega_p^2}{\omega_p^2 - \omega^2}. \tag{10.14}$$

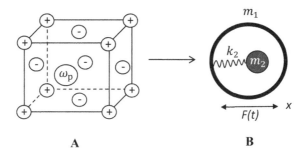

FIGURE 10.3 (A) Free electrons gas is embedded into the ionic lattice; ω_p is the electron plasma frequency (see Section 8.9). (B) The equivalent mechanical scheme of the system **A**. Core with mass m_2 (free electrons gas mass) is connected internally through the spring with $k_2 = \omega_p^2 m_2$ to a shell with mass m_1 (ionic lattice mass). The system is subjected to the time-dependent force $F(t) = F_0 \sin \omega t$.

It is clear from Eq. (10.14) that the effective dimensionless mass $\frac{m_{eff}}{m_1+m_2} \cong \frac{m_{eff}}{m_1}$ depends only on the ratio $\frac{m_2}{m_1}$; thus, it is independent on the metallic particles' size. Thus for the purposes of calculation, m_2 is taken as the mass of electron m_e and m_1 is the mass of the atom of metal.[5] The dependence of the dimensionless mass $m_{eff}/(m_1+m_2)$ on the dimensionless frequency ω/ω_p for two model metals Li and Au is reported in Reference 5. Again, the effective mass of the unit cell, depicted in Figure 10.3, is negative, when the frequency of external harmonic force ω approaches the plasma frequency from above, as it follows from Eq. (10.13).

10.3   NEGATIVE DENSITY METAMATERIALS

Now we are ready to "build" the negative density metamaterial. Recall that meta-materials are built from the macroscopic elementary bricks/units and not from the molecules (see Sections 9.3 and 9.4). Let us develop the 1D metamaterial built from the units depicted in Figure 10.2. The 1D lattice depicted in Figure 10.4 and studied in Reference 6 is built of identical elements possessing the effective negative masses m_{eff} given by Eq. (10.12) and connected by ideal springs k_1; the separation between springs a is constant. The effective density calculated in Reference 6 is given by Eq. (10.15):

$$\rho_{eff}(\omega) = \rho_{st} \frac{\theta}{\delta(1+\theta)\left(\frac{\omega}{\omega_0}\right)^2} \left\{ \cos^{-1}\left\{ 1 - \frac{\delta}{2\theta} \frac{\left(\frac{\omega}{\omega_0}\right)^2\left[\left(\frac{\omega}{\omega_0}\right)^2 - (1+\theta)\right]}{\left(\frac{\omega}{\omega_0}\right)^2 - 1} \right\} \right\}^2 , \quad (10.15)$$

where the static linear density of the chain ρ_{st} is given by $\rho_{st} = \frac{m_1+m_2}{a}$; $[\rho_{st}] = \frac{kg}{m}$ and $\theta = \frac{m_2}{m_1}$; $\delta = \frac{k_2}{k_1}$; $\omega_0 = \sqrt{\frac{k_2}{m_2}}$ and a is the lattice constant.[6] It was shown numerically in Reference 6 that the negative effective mass density $\rho_{eff}(\omega)$ occurs near the local

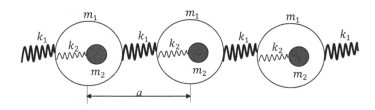

FIGURE 10.4 The mechanical scheme of the one-dimensional lattice giving rise to the negative effective density is depicted.

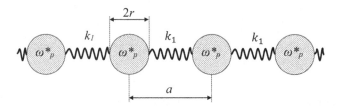

FIGURE 10.5 The 1D lattice exploiting plasma oscillations is depicted.

resonance frequency. This should be emphasized: the effect of the negative density emerges from the oscillations occurring in the vicinity of the resonance frequencies and is observed only close to these frequencies.

Now we consider the idea suggested in the previous section. Springs k_2 should not be necessarily mechanical springs; let us exploit the plasma oscillations instead of spring k_2 as suggested in the previous section and discussed in detail in References 5 and 7.

The 1D lattice exploiting plasma oscillations shown in Figure 10.5 will demonstrate the negative mass in the vicinity of the plasma frequency which is on the order of magnitude of $\omega_p \cong 10^{16}$ Hz, which is very high (see Section 8.9). However, this frequency may be decreased very strongly for meso-structures built of thin metallic wires, as discussed in Section 9.3 and demonstrated in References 8 and 9. Depression of the plasma frequency into the far infrared and even GHZ band becomes possible due to the mutual inductance appearing in the periodic arrays built of thin metallic wires (see Section 9.3).[8,9] A model of metamaterials built of Li and Au micro-particles embedded into polymer and glass matrices, represented by the referred ideal elastic springs with the constant k_1, was considered in Reference 7. The optical and acoustical branches of longitudinal modes propagating through the lattice are elucidated.[7] Acoustic negative-density materials were reported in Reference 10. The authors of Reference 10 exploited the alternative approach to the development of the negative density metamaterial, namely they reported the negative density material which is a tube with an array of thin tight membranes inside, which exhibits negative density below a cutoff frequency.

10.4   DOUBLE-NEGATIVE ACOUSTIC METAMATERIALS

First, let us introduce and define exactly the main notions of the classical theory of elasticity. Let us expose the cylindrical rod with the initial length l_0 and cross-section S to the stress σ, as shown in Figure 10.6. Young's modulus E describes the

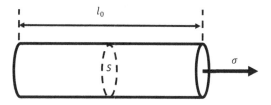

FIGURE 10.6   Uniaxial stretching of the cylindrical uniform rod by the stress σ is shown; the cross-section of the rode is S; the initial length of the rod is l_0.

relationship between stress (force per unit area $\sigma = \frac{F}{S}$) and strain, proportional relative deformation in an object $\varepsilon = \frac{\Delta l}{l_0}$, where Δl is the elongation of the rod. Young's modulus E is introduced according to Eq. (10.16):

$$\sigma = E\varepsilon. \tag{10.16}$$

Young's modulus is named for the 18th-century English physician and physicist Thomas Young. Young's modulus describes the elastic properties of a solid undergoing tension or compression in only one direction, as in the case of a metal rod that after being stretched or compressed lengthwise returns to its original length.[11] Young's modulus is a measure of the ability of a material to withstand changes in length when under lengthwise tension or compression.[11] Young's modulus is meaningful only in the range in which the stress is proportional to the strain (i.e., in the field of linear elasticity) and the material returns to its original dimensions when the external force is removed, in other words, when the deformation is elastic.[11] As stress increases, the material may either flow, undergoing permanent deformation, or finally break. In the engineering practice, the stretching testing of materials is reflected with the strain-stress curve. The typical stress-strain curve is shown in Figure 10.7. Young's modulus appears in the stress-strain curve as the slope of the linear portion of the curve (see Figure 10.7); generally speaking, the stress-strain curve includes the non-linear portions, as shown in Figure 10.7. These non-linear portions are out of the scope of the linear theory of elasticity.[11]

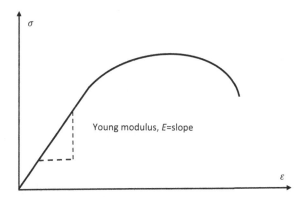

FIGURE 10.7   The stress-strain curve is shown; Young's modulus is the slope of the linear portion of the curve.

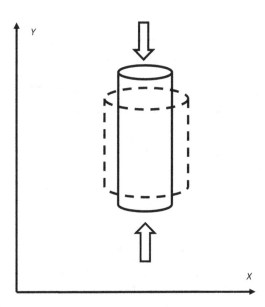

FIGURE 10.8   The Poisson effect is illustrated. A material tends to expand in directions perpendicular to the direction of compression; the compression is shown with arrows.

The Poisson ratio is a measure of the Poisson effect, the phenomenon in which a material tends to expand in directions perpendicular to the direction of compression, as shown in Figure 10.8, namely: the material is compressed in the y-direction and stretched in the x-direction.[11] Conversely, if the material is stretched rather than compressed, it usually tends to contract in the directions transverse to the direction of stretching. It is a common observation that when a rubber band is stretched, it becomes noticeably thinner. Assuming that the material is stretched or compressed in only one direction (see Figure 10.8), we define the Poisson ratio, usually denoted v, as follows:

$$v = -\frac{d\varepsilon_{trans}}{d\varepsilon_{axial}} = -\frac{d\varepsilon_y}{d\varepsilon_x} = -\frac{d\varepsilon_z}{d\varepsilon_x}, \tag{10.17}$$

where ε_{trans} and ε_{axial} are the transverse and axial strains, respectively. It should be emphasized that Eq. (10.17) is true for isotropic elastic materials. It is also important that for the isotropic elastic materials, the pair of parameters, namely Young's modulus and the Poisson ration, exhaust the elastic properties of the isotropic body.[11] In other words, any elastic deformation could be described with this pair of parameters. For example, the bulk modulus is expressed via Young's modulus and the Poisson ratio.

The bulk modulus of a substance (denoted K) is a measure of the resistance of a substance to bulk compression. It is defined as the ratio of the infinitesimal pressure increase to the resulting relative decrease of the volume. Consider the rectangular prism exposed to the uniform compression P, depicted in Figure 10.9. The

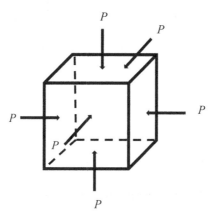

FIGURE 10.9   Rectangular prism exposed to the uniform compression P.

compression leads to the change in the volume of the prism dV. The bulk modulus is defined according to Eq. (10.18):

$$K = -V\frac{dP}{dV}.$$ (10.18)

The bulk modulus is expressed with Young's modulus and Poisson ration with Eq. (10.19):

$$K = \frac{E}{3(1-2v)}.$$ (10.19)

It is seen that the pair of parameters E and v define the bulk modulus K. Usually, the Poisson modulus is restricted within the range $0 < v < \frac{1}{2}$, which results in the positive values of the bulk modulus $K > 0$, which is quite expectable. Indeed, it is reasonable to suppose that the uniform compression, illustrated with Figure 10.9, will result in the decrease of the volume of the compressed body. However, there exist materials for which the bulk modulus is negative, and $K < 0$ is true. These materials are called auxetic materials (or auxetics). It is interesting that conventional materials (built of molecules) such as α-cristobalite demonstrate negative values of the Poisson ratio.[12]

It was suggested that the experimentally observed negative Poisson ratios in the silicate α-cristobalite may be related a two-dimensional "rotation of rigid units" model involving "rotating rectangles".[12] However, a diversity of auxetics were developed as metamaterials.[13-15] These auxetic materials include honeycomb materials, foams and microporous polymer cylinders.[13-15]

Now, we address the double-negative acoustic materials (abbreviated DNG), which are analogue of electromagnetic DNG materials discussed in much detail in the previous chapter (Sections 9.1 and 9.3). In DNG acoustic materials, the Poynting vector $\vec{S}$ (see Section 7.1.11) and the wave vector $\vec{k}$ (see Section 7.1.8) should point in the opposite directions.[16] Double negative in acoustics implies simultaneous negativity in bulk modulus, supplied by Eq. (10.19) and density.[16] The double negativity in acoustics is derived from low-frequency resonances, as in the case of electromagnetism, but the negative density and modulus are derived from a single resonance structure as distinct from electromagnetism in which the negative permeability and negative permittivity originate from different resonance mechanisms.[16]

The explanation of the phenomena of negative density and bulk modulus, based on the concepts of hidden force and hidden source of volume, is suggested in Reference 16. Hidden force and hidden source of volume are responsible for the effective densities and moduli of acoustic metamaterials.[16] The superficially strange concepts of negative density and modulus are naturally accounted for in this picture when the hidden force/source operates in antiphase to and is bigger in magnitude than the force/volume-change engendered in its absence. DNG air-based metamaterials involving membranes, Helmholtz resonators or side holes with the inclusion were discussed in detail in Reference 17. Sound manipulation with metamaterials and acoustic cloaking is discussed in References 18–22 Negative refraction of sound by the metamaterial built of porous rubber microbeads suspended in a gel was reported in Reference 23, which makes these metamaterials promising for underwater acoustic applications. Acoustic metamaterials demonstrate an essential potential for noise-reduction.[24]

Bullets

- Mechanical systems in which the direction of the external force and acceleration of the body are oriented in opposite directions are possible. These systems give rise to the effect of the "negative effective mass".
- Demonstrating negative effective mass includes the hidden elements built of oscillating masses. The oscillations result in the "negative effective mass". The negative effective mass is a resonant effect and it is observed in a vicinity of the fixed resonance frequencies.
- Metamaterials built of the elements/units demonstrating the negative effective mass show the "negative effective density". The "negative effective density" is the resonant effect.
- The negative effective density may be observed in the vicinity of the plasma frequency.
- Main notions of the theory of linear elasticity are introduced.
- Auxetic materials demonstrating negative bulk modulus are introduced and discussed.
- Double-negative acoustic materials demonstrating simultaneous negativity in density and bulk modulus are discussed.
- Applications of acoustic metamaterials are addressed.

EXERCISES

1. Demonstrate Eq. (10.11).
2. What is the physical meaning of the notion of the negative effective mass, emerging from Eqs. (10.11) and (10.12)?

    Answer: The negative effective mass means that the force acting on the body depicted the acceleration of the body is oriented in the opposite directions.

3. How the value of the plasma frequencies may be decreased?

   Hint: The plasma frequency may be decreased periodic structures built of thin metallic wires, as discussed in Section 9.3 (see Figure 9.5).

4. Explain the meaning of the Poisson ratio v. What are the dimensions of the Poisson ratio?

5. What is the physical meaning of the bulk modulus K? What are the dimensions of the bulk modulus?

6. What is the meaning of the notion "auxetic materials"? Supply the examples of auxetic materials.

7. Derive Eq. (10.19).

8. What are the relative directions of the Poynting vector $\vec{S}$ and the wave vector $\vec{k}$ in the double-negative acoustic metamaterials?

9. What are the possible applications of acoustic metamaterials (see Reference 18)?

10. Explain the effect of the acoustic cloaking with the acoustic metamaterials.

REFERENCES

1. Sheng P., Zhang X. X., Liu Z., Chan C. T. Locally resonant sonic materials. *Phys. B. Condens. Matter* 2003, **338**, 201–205.
2. Liu Z., Chan C. T., Sheng P. Analytic model of phononic crystals with local resonances. *Phys. Rev. B* 2005, **71**, 014–103.
3. Chan C. T., Li J., Fung K. H. On extending the concept of double negativity to acoustic waves. *J. Zhejiang Univ. A* 2006, **7**, 24–28.
4. Milton G. M., Willis G. R. On modifications of Newton's second law and linear continuum elastodynamics. *Proc. R. Soc. A* 2007, **463**, 855–880.
5. Bormashenko E., Legchenkova I. Negative effective mass in plasmonic systems. *Materials* 2020, **13** (8), 1890.
6. Huang H. H., Sun C. T., Huang G. L. On the negative effective mass density in acoustic metamaterials. *Int. J. Eng. Sci.* 2009, **47**, 610–617.
7. Bormashenko E., Legchenkova, Frenkel M. Negative effective mass in plasmonic systems II: Elucidating the optical and acoustical branches of vibrations and the possibility of anti-resonance propagation. *Materials* 2020, **13** (16), 3512.
8. Pendry J. B., Holden A. J., Stewart W. J., Youngs I. Extremely low frequency plasmons in metallic mesostructures. *Phys. Rev. Lett.* 1996, **76**, 4773.
9. Pendry J. B. Negative refraction. *Contemp. Phys.* 2004, **45** (3), 191–202.
10. Lee S. H., Park C. M., Seo Y. M., Wang Z. G., Kim C. K. Acoustic metamaterial with negative density. *Phys. Lett. A* 2009, **373**, 4464–4469.
11. Gould P. L., Feng Y., *Introduction to Linear Elasticity, 4th Ed.*, Chapters 2–4, Springer, New York, USA, 2018.
12. Grima J. N., Gatt R., Alderson A., Evans K. E. On the origin of auxetic behavior in the silicate α-cristobalite. *J. Mater. Chem.* 2005, **15**, 4003–4005.
13. Alderson A., Alderson K. L. Auxetic materials. *Proc. Inst. Mech. Eng. Part G: J. Aerosp. Eng.* 2007, **221**, 565–575.
14. Yang W., Li Z. M., Shi W. Review on auxetic materials. *J. Mater. Sci.* 2004, **39**, 3269–3279.
15. Joseph A., Mahesh V., Harursampath D. On the application of additive manufacturing methods for auxetic structures: A review. *Adv. Manuf.* 2021, **9**, 342–368.
16. Li J., Chan C. T. Double-negative acoustic metamaterial. *Phys. Rev. E* 2004, **70**, 055602(R).
17. Lee S. H., Wright O. B. Origin of negative density and modulus in acoustic metamaterials. *Phys. Rev. B* 2016, **93**, 024302.

18. Cummer S., Christensen J., Alù A. Controlling sound with acoustic metamaterials. *Nat. Rev. Mater.* 2016, **1**, 16001.
19. Ma G., Sheng P. Acoustic metamaterials: From local resonances to broad horizons. *Sci. Adv.* 2016, **2** (2), e1501595.
20. Li J., Wen X., Sheng P. Acoustic metamaterials. *J. Appl. Phys.* 2021, **129**, 171103.
21. Ma L., Cheng L. Sound radiation and transonic boundaries of a plate with an acoustic black hole. *J. Acoust. Soc. Am.* 2019, **145**, 164–172.
22. Wu Y., Yang M., Sheng P. Perspective: Acoustic metamaterials in transition. *J. Appl. Phys.* 2018, **123**, 090901.
23. Popa B. I., Cummer S. Negative refraction of sound. *Nature Mater.* 2015, **14**, 363–364.
24. Gao N., Zhang Z., Deng J., Guo X., Cheng B., Hou H. Acoustic metamaterials for noise reduction: A review. *Adv. Mater. Technol.* 2022, **7** (6), 2100698.

Index

A

Adhesives, 77–78, 91
Advancing contact angle, 34–36
Apparent contact angle, 37

B

Bond number, 13
Bulk modulus, 176–177

C

Capillarity, 32
Capillary rise, 31–33
Cassie wetting, 41–44
Cassie-Baxter equation, 41–42
Complete wetting, 24
Contact angle hysteresis, 34–38
Cox-Voinov law, 39

D

Damped oscillations, 101–103
Debye interaction, 8, 87
Derjaguin isotherm, 30
Dielectric permittivity, 150, 163
Disjoining pressure, 27–30
Dispersion (London force), 8–9, 88–89
Dispersion relation, 109
Double-negative material, 163–166
Drude model, 142–146
Dry adhesives, 91
Dynamic contact angle, 38–39

E

Electromagnetic cloaking, 159–160
Electromagnetic waves, 116–121

F

Fermat principle, 133–136, 166

G

Gecko effect, 77–91
Group velocity, 112–116

H

Hamaker constant, 27–28
Harmonic oscillations, 94–101

J

Johnson–Kendall–Roberts theory, 78–79

K

Keesom interaction, 8–9, 88–89

L

Laplace pressure, 11–13
Left handed materials, 156–158
Lens, 158–159
Line tension, 25–27
Longitudinal waves, 109–112

M

Magnetic meta-materials, 162–165
Magnetic permeability, 150, 163–165
Maxwell equations, 116–117
Meta-materials, 156–166, 173–178

N

Negative density, 173–174
Negative mass, 169–173
Negative refractive index, 155–160

P

Phase velocity, 112–116, 119
Plasma frequency, 146–150
Plasma oscillations, 146–147
Poisson ratio, 176–177
Poynting vector, 121

R

Receding contact angle, 34–35
Refractive index, 128–133, 136–142, 147–150, 155–161
Resonance, 104–106
Reynolds number, 17, 73
Right handed materials, 156–157
Rose petal effect, 65–67
Rough surface, 39–40

S

Salvinia effect, 68–69
Shark skin effect, 72–75
Snell law, 129

Printed in the United States
by Baker & Taylor Publisher Services